Smooth Tests of Goodness of Fit

Smooth Tests of Goodness of Fit

J. C. W. RAYNER

Department of Mathematics and Statistics, University of Otago, Dunedin, New Zealand

D. J. BEST

CSIRO, IAPP Biometrics Unit, Food Research Laboratory, North Ryde, Australia

New York Oxford
OXFORD UNIVERSITY PRESS
1989

Oxford University Press

Oxford New York Toronto
Delhi Bombay Calcutta Madras Karachi
Petaling Jaya Singapore Hong Kong Tokyo
Nairobi Dar es Salaam Cape Town
Melbourne Auckland

and associated companies in
Berlin Ibadan

Published by Oxford University Press, Inc.,
200 Madison Avenue, New York, New York 10016

Oxford is a registered trademark of Oxford University Press

Library of Congress Cataloging-in-Publication Data
Rayner, J. C. W.
Smooth tests of goodness of fit.
Bibliography: p. Includes index.
1. Goodness-of-fit tests. I. Best, D. J.
II. Title.
QA277.R395 1989 519.5′6 88-34489
ISBN 0-19-505610-8

1 2 3 4 5 6 7 8 9

Printed in the United States of America
on acid-free paper

J. C. W. R.: To the Rayners, Fred, Carol, Glen and Eric. And my friend and colleague the first fleeter.

D. J. B.: To my parents and to Helen, Warwick and Rohan.

Preface

The importance of probabilistic or statistical modeling in the modern world cannot be overrated. With the advent of high-speed computers, complex models for important processes can now be constructed and implemented. These models and the associated statistical analyses are of great assistance in making decisions in diverse fields, from marketing, medicine, and management, to politics, weapons systems, and food science. *Goodness of fit* is concerned with assessing the validity of models involving statistical distributions, an essential and sometimes forgotten aspect of the modeling exercise. One can only speculate on how many wrong decisions are made due to the use of an incorrect model.

Karl Pearson pioneered goodness of fit in 1900, when his paper introducing the X^2 test appeared. Since then, perhaps reflecting the needs and importance of the subject, a great many new tests have been constructed. The *smooth tests* are a class of goodness of fit tests that are informative, easy to apply, and generally applicable. Typically they can be derived as score tests, and hence are, in a sense, optimal for large sample sizes. For moderate and small samples they are very competitive in the cases we have examined. Pearson's X^2 test is in fact a smooth test. In the formulation we prefer, components with simple graphic interpretations are readily available. We suggest that the properties of the smooth tests are such that a new goodness of fit test must be in some way superior to a corresponding smooth test if it is to be worthy of consideration.

This book is complementary to that by D'Agostino and Stephens (*Goodness of Fit Techniques,* 1986) in that they do not cover the smooth tests in any detail, while we do not cover in detail topics such as tests based on the empirical distribution function, and tests based on regression and correlation. There is some overlap in the coverage of X^2 tests. The tests that they discuss and our smooth tests are in competition with each other. We give some comparisons, and, not surprisingly, recommend use of the smooth tests. Usually, the smooth tests are more informative than their competitors. The D'Agostino and Stephens book covers a broad range of topics, generally omitting mathematical details and including many tables and examples so that it reads as a handbook of methods. Since our book concentrates on smooth methods, we have been able to present derivations and mathematical details that might have been omitted in a more comprehensive treatment of goodness of fit in its entirely. We consider this to be highly desirable because the development of the smooth tests of fit is far from complete. Indeed, we hope that researchers will read this book and be motivated to help with its further development.

In spite of this last statement, many economists, scientists and engineers who have taken an introductory mathematical statistics course, to the level of Larsen and Marx (1981), will be able to read this book. The more technical details are clearly signposted and are in sections that may be omitted or skimmed. Undergraduates with a sufficient background in statistics and calculus should be able to absorb almost everything. Practical examples are given to illustrate use of the techniques.

The smooth tests for the uniform distribution were introduced by Neyman (1937), but they were slow to gain acceptance because the computations are heavy by hand. This is no longer a barrier. Many of the techniques we discuss are readily implemented on modern computers, and we give some algorithms to assist in doing this. When used in conjunction with density estimate plots or Q–Q plots, the smooth tests can play an important part in many analyses.

Chapter 1 outlines the goodness of fit problem, gives a brief history of the smooth tests, outlines the monograph , and gives some examples of the sort of problems that arise in practice. A review of Pearson (1900), and an outline of the early development of the tests for simple and composite hypotheses is given in Chapter 2. In Chapter 3, tests that are asymptotically optimal are introduced; these include the score tests that are particularly important later in the book. Using score tests and smooth models, tests of completely specified null hypotheses are derived in Chapters 4 and 5. These chapters cover both uncategorized (discrete or continuous) and categorized the null distributions. The tests are essentially tests for uniformity. Then, in Chapters 6 and 7, we consider tests for composite null hypotheses, again treating both the categorized and uncategorized cases. Chapters 4 to 7 emphasize the components our tests yield. In Chapter 6 we look at tests for the univariate and later for the multivariate normal, the Poisson, geometric and exponential distributions. These are extracted from a class of smooth goodness of fit tests. In Chapter 7, we discuss X^2 statistics for composite hypotheses. We conclude with a review and an examination of some of the other uses to which our techniques may be put.

Our interest in the subject of goodness of fit came about partly from questions from colleagues relating to the "Gaussian" assumption in routine statistical analyses, and partly from work J. C. W. R. had begun as a student of H. O. Lancaster. Our approach is based on the use of orthonormal functions, emphasized in Lancaster (*The Chi-Squared*

Distribution, 1969), and on the use of score statistics and generalizations of the smooth families suggested by Thomas and Pierce (1979) and Kopecky and Pierce (1979) in articles published in *The Journal of the American Statistical Association*.

Dunedin, New Zealand J. C. W. R.
North Ryde, Australia D. J. B.

Acknowledgments

We would like to thank many friends and colleagues for their generous contributions, ranging from encouragement and moral support, to proof-reading, typing, and technical advice. In particular, we would like to thank, in alphabetical order, Mary-Jane Campbell, Peter Diggle, Geoff Eagleson, Lynette Grant, Derek Holton, Fred Lam, Bryan Manly, Lyn McAlevey, Tony Pettitt, Allyson Seyb, and Terry Speed.

In addition a grant made by the Otago University Research Committee is gratefully acknowledged.

Contents

Smooth Tests of Goodness of Fit

1

Introduction

1.1 The Problem Defined

A number of statistical or probabilistic distribution models have found widespread use in scientific, economic, medical, and engineering fields. This monograph is concerned with assessing the goodness of fit of such models, or seeing how well the data agrees with the proposed distribution model.

One informal way of doing this is to draw a histogram of the frequency distribution of the data and judge the closeness of the histogram and the theoretical probability function "by eye." Such graphic checks are subjective and cannot be recommended to assess goodness of fit *on their own*. The hypothesis tests we recommend may be complemented by graphic methods.

Examples of the use of the models we discuss are many. Almost all of the commonly used methods of statistical inference, such as t-tests, determination of P values and least significant differences, assume a normal or Gaussian distribution. The method of least squares is a common estimation technique but has optimum properties only when normality is assumed. More specific examples are as follows:

1. Quality control rules for the import and export of various foods often assume a normal distribution,
2. Safety limits for extreme rainfall used by hydrologists involved in flood control may assume a lognormal distribution,
3. Estimates of bacteria in sewage may be based on an exponential distribution.

Failure of the distributional assumptions means failure of the model. The conclusions based on the model are then invalid. This possibility of failure can be assessed objectively by a goodness of fit test.

A number of previous books have recognized the importance of goodness of fit tests by devoting a section or chapter to the topic. For example, see Tiku et al. (1986, Chapter 6), Chambers et al. (1983, Chapter 6), Lawless

(1982, Chapter 9), Shapiro and Gross (1981, Chapter 6), Gnanadesikan (1977, Chapter 5), and Kendall and Stuart (1973, Chapter 30). These treatments, however, are somewhat limited in scope. D'Agostino and Stephens (1986) gave goodness of fit a much more comprehensive coverage. However, none offer the coherent approach of this monograph, which concentrates on one class of tests and develops both theory and applications. This monograph also gives more coverage to categorized models.

Without a statistical test, goodness of fit can only be assessed by visual subjective methods. R. A. Fisher (1925) in his influential text, *Statistical Methods for Research Workers,* devoted large sections of his Chapters 3 and 4 to goodness of fit and commented on the then common model assessment method:

> No eye observation of such diagrams, however experienced, is really capable of discriminating whether or not the observations differ from the expectation by more than we would expect from the circumstances of random sampling. (R. A. Fisher, 1925, p. 36)

Kempthorne (1966) considered goodness of fit to be the "classical problem of statistical inference." What then is a goodness of fit test? According to David (1966, p. 399),

> A goodness of fit procedure is a statistical test of a hypothesis that the sampled population is distributed in a specific way . . . for example, that the sampled population is normal.

This is the one sample problem: the corresponding k-sample problem assesses whether or not k independent random samples come from the same popultion.

Subsequently, we shall mainly be concerned with one sample tests for goodness of fit. Formally, given a random sample $X_1, X_2, \ldots, X_n$, we test the null hypothesis that the sampled population has cumulative distribution function $F(x; \theta)$, $\theta \in \Theta$, against the alternative hypothesis that the cumulative distribution function is $G(x; \omega)$, $\omega \in \Omega$. All of X, Θ, and Ω may be multidimensional. Frequently, the alternative is simply "not the null hypothesis."

What do we get from having applied a goodness of fit test? First, a *compact description* of the data. Saying that the data is binomial with parameters $n = 15$ and $p = .51$ is a valuable abbreviation of the available information. Second, *powerful parametric procedures,* such as the tests in the analysis of variance, are valid if the data are consistent with normality. And third, *light may be shed on the mechanisms generating the data.* For example, if the data cannot be viewed as a Poisson process, then at least one of the axioms sufficient for a Poisson process has failed. If lifetimes for cancer patients from the onset of "standard" treatment have been exponentially distributed with a mean of 36 months in the past, and this distribution no longer holds under a new treatment, what has changed? It could be that either the mean or the distribution has changed. In the latter case perhaps treatment is less effective than the standard treatment for some, and they die sooner than under the standard treatment; conversely, the treatment is apparently effective for others, who survive longer than previously.

What a goodness of fit test tells us is important, but so is what it doesn't tell us! Geary (1947) said that

> Normality is a myth; there never was, and never will, be a normal distribution.

Strongly put perhaps, but given enough observations of virtually any generating mechanism, we could probably reject any specified hypothesis. As for normality, we do not observe arbitrarily large (or small) data; and as all data are rounded, we should ultimately be able to reject any continuous model. But although a distributional model may not hold precisely, it may hold sufficiently well for the three purposes outlined earlier. The important question is, *are our data sufficiently well approximated by the distribution for which we test*?

Many data sets are summarized by a statement of the form "mean ± standard error." This *assumes* a normal distribution, or at least a distribution that is completely determined by the mean and standard deviation. If the data were thought to be Poisson, then it would be sufficient simply to quote the mean. But of course in such cases the distribution should be assessed by a goodness of fit test.

Should the common tests of statistical inference, such as the t-test and the analysis of variance, be avoided by the use of more robust, distribution-free or nonparameteric procedures? The latter minimize distributional assumptions and, at times, this minimization is a wise course to follow. In some cases, however, not using parametric tests can result in the use of inferior tests. We suggest that goodness of fit tests and other checks on the data should be employed before opting for robust or distribution-free techniques.

This leads to a difficulty. If a preliminary test of the assumptions for a parametric test is performed, does this affect the inferences made? We agree with Cox (1977) who proposed:

> A combination of preliminary inspection of the data together with study at the end of the analysis of whether there are aspects of the data and assumptions reconsideration of which might change the qualitative conclusions.

In our circumstances, we interpret this as meaning the parametric test is inapplicable if the distributional assumptions are not satisfied, so there is no need to incorporate the results of a goodness of fit test formally. The fact that a goodness of fit test is applied formally, does not mean it is not, under some circumstances, part of the preliminary inspection of the data.

A distinction can be drawn between *omnibus* and *directional* tests. Omnibus tests are intended to have moderate power against all alternatives; directional tests are intended to detect specified alternatives well. Of course, against the specified alternatives, the directional tests are constructed to be more powerful than the omnibus tests, while against all other alternatives the omnibus tests should be superior. Consider the analogy of several torches shining in one direction in the dark, and therefore lighting almost everything in that direction, but almost nothing in any other direction. Compared to this, consider the torches shining in several different directions and therefore giving fair discrimination in all directions. The smooth tests are constructed to be onmibus tests, but their *components* provide powerful directional tests.

Finally, we mainly discuss formal statistical tests of significance. This is not to say that subjective methods are not valuable. Graphic methods may lead to insights that are not apparent otherwise, and methods such as quantile–quantile (Q–Q) plots or density estimates should be used along with those we discuss there.

Now we turn to a brief history of smooth goodness of fit tests.

1.2 A Brief History of Smooth Tests

Perhaps the most widely known test in statistical inference is Pearson's X^2 goodness of fit test. An informal definition follows. Suppose observations may fall into m nonoverlapping classes or cells. We hypothesize the cells should contain respectively $E_1, \ldots, E_m$ observations, but the observed cell counts are $O_1, \ldots, O_m$. Now define the Pearson test statistic by

$$X_P^2 = \sum_{i=1}^{m} (O_i - E_i)^2 / E_i$$

If this is larger than the $100\alpha\%$ point of the χ^2_{m-1} distribution then the hypothesized expectations can be rejected at the $100\alpha\%$ level of significance. The test is more formally defined in §2.2. Note that we used the following terminology. Pearson's X^2 test uses the statistic X_P^2 which has asymptotic distribution χ^2_{m-1}. It reduces ambiguity if the test is identified by its test statistic rather than the statistic's (asymptotic) distribution. Since the symbol X_P^2 is not universal, we usually say Pearson's X^2 test, or just Pearson's test.

Pearson's test is applicable for testing discrete data when there are no parameters that need to be estimated. The expansion of the methodology to cover more practical situations has occupied statisticians almost continuously since Karl Pearson introduced his X^2 test in 1900. In the next chapter we will devote some time to reviewing Pearson (1900) and the developments in X^2-type tests. It is not widely known that Pearson's test is a smooth test, but we will demonstrate that this is the case in Chapter 5.

According to Barton (1956) and Neyman himself (1937), Neyman's (1937) smooth test was developed to overcome presumed deficiencies in Pearson's X^2 test. The test was called "smooth" because it was constructed to have good power against alternatives whose probability density functions depart "smoothly" from that specified by the null hypothesis. For example, the null hypothesis may specify the normal distribution with zero mean and unit variance, while the alternative may specify the normal distribution with unit mean and unit variance. Smooth changes include shifts in mean, variance, skewness, and kurtosis.

Suppose we have a random sample from a continuous distribution with completely specified cumulative distribution function $F(x)$. Applying the probability integral transformation, the null hypothesis H_0 specifies that $Y = F(X)$ is uniformly distributed on (0, 1). Neyman's smooth *alternative of order k* to H_0 has probability density function

$$g_k(y;\theta) = \exp\left\{\sum_{i=1}^{k} \theta_i \pi_i(y) - K(\theta)\right\}, \qquad 0 < y < 1; \quad k = 1, 2, 3, \ldots \qquad (1.2.1)$$

where $\theta^T = (\theta_1, \ldots, \theta_k)$, $K(\theta)$ is a normalizing constant, and the $\pi_i(y)$ are orthonormal polynomials related to the Legendre polynomials. The first five such polynomials are:

$$\pi_0(y) = 1, \qquad \pi_1(y) = \sqrt{3}(2y-1), \qquad \pi_2(y) = \sqrt{5}\,(6y^2 - 6y + 1)$$
$$\pi_3(y) = \sqrt{7}(20y^3 - 30y^2 + 12y - 1)$$
$$\pi_4(y) = 3(70y^4 - 140y^3 + 90y^2 - 20y + 1)$$

The $\pi_i(y)$ are constructed so that $\pi_i(y)$ is of degree i and the $\pi_i(y)$ are orthonormal on (0, 1) (for example, see Kendall and Stuart, 1973, p. 444). To test the null hypothesis H_0: $\theta_1 = \ldots = \theta_k = 0$, we use the Neyman statistic, given by

$$\Psi_k^2 = \sum_{i=1}^{k} U_i^2, \quad \text{in which} \quad U_i = \sum_{j=1}^{n} \pi_i(Y_j)/\sqrt{n}$$

The U_i are called *components* of Ψ_k^2.

Neyman's conception for his smooth test was that it should be constructed to be locally most powerful, unbiased, and of size α for testing for uniformity against the order k alternatives given by Equation 1.2.1. Its power function was also constrained to be symmetric, depending on θ only through $\theta_1^2 + \ldots + \theta_k^2$. Neyman (1937) noted that his solution is only approximate; the test is only of size α, unbiased, and most powerful asymptotically. A detailed account of Neyman (1937) is given in §4.1.

Barton (1953, 1955, 1956) extended Neyman's work. He used probability density functions asumptotically equivalent to $g_k(y;\theta)$. For example, in Barton (1953) he used the probability density functions

$$h(y;\theta) = 1 + \sum_{i=1}^{k} \theta_i \pi_i(y), \qquad 0 < y < 1; \qquad k = 1, 2, 3, \ldots \qquad (1.2.2)$$

His 1956 paper dealt with probability density functions involving nuisance parameters, but the statistic derived had an inconvenient distribution. As Kopecky and Pierce (1979) pointed out, the quadratic score statistic (see Chapter 3) has a more convenient distribution.

An interesting but little known result is that the Pearson X^2 test is a categorized form of the Neyman–Barton tests. Suppose a multinomial with

g classes is specified by the null hypothesis. Barton (1955) considered order k alternatives of the form $h(y; \theta)$, but with the polynomials $\pi_i(y)$ replaced by an orthonormal system on the multinomial distribution. He then defined a statistic $B(g, k)$ that as $g \to \infty$, (1) approached Ψ_k^2, (2) tended to be distributed as χ_k^2, and (3) was optimal in the limit. Moreover $B(k+1, k)$ was shown to be equivalent to the Pearson test statistic based on $k+1$ classes. The importance of this result is that the Pearson X^2 test with $k+1$ cells can be expected to have good power properties against order k alternatives, especially for a moderate to large number of classes, when it will be very similar to the optimal Ψ_k^2. Kendall and Stuart (1973, p. 446) reviewed this material and showed that the $B(g, k)$ may be obtained by partitioning the Pearson test statistic. This idea will be taken up again in Chapter 4.

Watson (1959) extended a result of Barton (1956), and Hamdan (1962, 1963, 1964), considered smooth tests for various simple null hypotheses. He used the Hermite–Chebyshev polynomials to construct a test for the standard normal distribution, and an orthonormal set on the multinomial and the Walsh functions to construct tests for the uniform distribution.

These tests aroused little interest. They required computations that would be considered heavy by hand, and could not deal practically with the main interest in applications, composite null hypotheses. It was not until the papers of Thomas and Pierce (1979) and Kopecky and Pierce (1979) that Neyman-type tests received much further attention.

Rather than work with orthogonal polynomials, Thomas and Pierce (1979) defined an order k probability density function by

$$\exp\left\{\sum_{i=1}^{k} \theta_i y^i - K(\theta)\right\}$$

or, in terms of the null probability density function,

$$\exp\left\{\sum_{i=1}^{k} \theta_i F^i(x) - K(\theta)\right\} f(x)$$

where $f(x) = dF(x)/dx$. Their test statistic W_k^* is a quadratic score statistic based on this model. The weak optimality of tests based on the quadratic score statistics is therefore conferred upon the W_k^*.

If the probability density function $f(x)$ involves nuisance parameters, the model for an order k alternative becomes

$$\exp\left\{\sum_{i=1}^{k} \theta_i F^i(x; \beta) - K(\theta)\right\} f(x; \beta)$$

The quadratic score statistic based on this model is W_k, given in detail in Thomas and Pierce (1979, p. 443), and the tests based on the W_k can be interpreted as generalized smooth tests.

Write $\beta = (\mu, \sigma)^T$, and write $F(x; \beta)$ for the cumulative distribution function of the normal distribution with unknown mean (μ) and unknown variance (σ^2). In testing for this distribution Thomas and Pierce suggested the statistics

$$W_1 = \left\{16.3172 \sum (Y_j - 0.5)\right\}^2 \Big/ n$$

and

$$W_2 = W_1 + 27.3809^2 \left\{\sum (Y_j^2 - 1/3) - \sum (Y_j - 0.5)\right\}^2 \Big/ n$$

where $Y_j = F(X_j; \hat{\beta})$. The statistics W_1 and W_2 are asymptotically distributed as χ_1^2 and χ_2^2 respectively. Thomas and Pierce (1979) showed that the small sample distributions are reasonably approximated by the limiting χ^2 distributions.

Tests based on what might be called the Pierce approach include Bargal and Thomas' (1983) test for the (censored) Weibull, and Bargal's (1986) test for the (censored) gamma. Unfortunately, a consequence of using powers instead of orthonormal functions is that tables of constants, such as 16.3172 and 27.3809 in W_1 and W_2, are needed to define the tests statistics. This is somewhat offset by the need to know the orthonormal functions in the formulation we prefer. Fortunately, the orthonormal functions required for the more frequently occurring problems may be obtained from recurrence relations. This is most convenient for computer implementation of the tests. We will give some recurrence relations in Appendix 1.

Rayner and Best (1986), Koziol (1986, 1987), and Jacque and Bera (1987) all suggested smooth tests for the composite case when the parameters are of location-scale type. Their tests are based on orthonormal functions and are of slightly simpler form than those of Thomas and Pierce (1979) in that

1. They involve sums of squares and not quadratic forms.
2. Numerical integration is not usually needed to specify constants in the test statistic.
3. The components are often identifiable with known moment-type statistics used in tests of fit.
4. The components are asymptotically independent.

In particular Rayner and Best (1986) combined the Neyman and Pierce approaches to produce smooth tests for location-scale families. The order k alternative has the form

$$g^*(y; \theta) = \exp\left\{\sum_{i=1}^{k} \theta_i H_i(y; \beta) - K(\theta)\right\} f(y; \beta), \qquad k = 1, 2, 3, \ldots$$

and the appropriate test statistic is

$$\hat{S}_k = \sum \hat{V}_i^2, \quad \text{where} \quad \hat{V}_i = \sum H_i\{(X_j - \hat{\mu})/\hat{\sigma}\}/\sqrt{n}$$

in which the $H_i(z)$ are orthonormal on the standardized probability density function, $\hat{\mu}$ and $\hat{\sigma}$ are maximum likelihood estimates of μ and σ, respectively, and the summation is over nonzero summands. For example, in testing for the normal distribution with unspecified mean μ and variance

σ^2, the first two $\hat{V}_i$'s are zero; the next two assess skewness and kurtosis. The orthonormal functions in this case are the normalized Hermite polynomials, and those of order 3 to 6 are:

$$H_3(x) = (x^3 - 3x)/\sqrt{6},$$

$$H_4(x) = (x^4 - 6x^2 + 3)/\sqrt{24},$$

$$H_5(x) = (x^5 - 10x^3 + 15x)/\sqrt{120},$$

$$H_6(x) = (x^6 - 15x^4 + 45x^2 - 15)/\sqrt{720}$$

The test statistic and the components $\hat{V}_i^2$ all have asymptotic χ^2 distributions, so the test is easy to implement. Moreover, the components may be individually informative. Examples are given in Best and Rayner (1985b). This approach can readily be extended beyond location-scale families, and to categorized data and multivariate distributions. This will be done in later chapters.

To complete the picture, we should note that a smooth model appropriate in the discrete case, and in the absence of nuisance parameters, is

$$\pi_j = C(\theta) \exp\left\{\sum_{j=1}^{k} \theta_i h_{ij}\right\} p_j, \qquad j = 1, \ldots, m$$

Here, k is the order of the alternative, and the Pearson X^2 test results if in the score statistic we take $k = m - 1$ and choose the h_{ij} appropriately. The appeal in this formulation is that if the null hypothesis is rejected, with significant components indicating particular θ_i positive, then $\{\pi_j\}$ from the preceding equation specifies an alternative model. Lack of a suggested alternative hypothesis has been a criticism of X^2-type tests. The formulation here, and its composite analogue, will also be investigated later in Chapters 5 and 7.

1.3 Monograph Outline

The reader is now acquainted with what goodness of fit tests are, and why they are important. We have sketched the historical development of the smooth tests, and in future chapters we will return to that development in more detail.

In Chapter 2 we will begin at the chronological beginning, with Pearson's X^2 test. A review of Pearson (1900) is given, and also of the developments in X^2-type tests since then. This is not done from the viewpoint of smooth tests, since they were a later development. Certain X^2-type tests are smooth tests, as we have already mentioned. This will be demonstrated in Chapters 5 and 7.

The main approach will be to define smooth models in various situations, and to derive tests that have good properties in large samples for these models. The machinery for doing this is given in Chapter 3, on asymptotically optimal tests. The likelihood–ratio, score, and Wald tests are introduced for models first without, and second with, nuisance parameters. These tests are asymptotically equivalent, and which is most convenient to apply will vary depending on the situation. In multivariate normal models it is usually the likelihood–ratio test. For the smooth models we discuss, it is usually the score test.

Chapters 4 to 7 systematically work through derivations and properties of the score tests for *categorized* and *uncategorized* smooth models of the Neyman type, both when *nuisance parameters are absent* and when they are *present.* In this monograph uncategorized models will be either discrete or continuous, whereas the data are placed into a finite number of cells of classes in categorized models. This involves some ambiguity. For example the binomial could be treated as either categorized or uncategorized. Particular cases of the tests we derive include tests for the univariate and multivariate normal, exponential, geometric, and Poisson distributions. Power studies are given to demonstrate the effectiveness of these tests in small samples. The tests are applied to real data sets.

Throughout the monograph we will need to calculate (approximate) *P values* for certain data sets. Given a value of a test statistic for a data set, we usually need the probability of values of the test statistic at least as great as the observed under the null hypothesis. This is done by simulating a large number (at least 200, preferably at least 1,000) of random samples from the null distribution. The proportion of these samples for which the value of the test statistic is at least the observed is calculated. This is the P value. It is approximately the probability of observations at least as extreme as the observed statistic under the null hypothesis.

Calculation of P values requires the generation of random deviates. Most computers used in scientific or engineering applications provide a random uniform $(0, 1)$ generator. To produce the random normal deviates needed for the approximate P values, the simple Kinderman and Monahan (1977) algorithm was used. This is

1. Generate random uniform $(0, 1)$ variates u_1 and u_2. Let $x = 1.715528(u_1 - 0.5)/u_2$ and $z = x^2/4$.
2. If $z \leq -\ln(u_2)$ accept x as a random $N(0, 1)$ value; otherwise return to step 1.

Best (1979) showed this algorithm to be reasonably efficient. Devroye (1986) gave details of algorithms for the generation of random deviates for most of the common distributions.

Throughout the monograph we will freely use graphic methods to augment our tests. It would be inappropriate for us to expound their virtues, so the reader is instead directed to other sources. For example, see Chambers et al. (1983) and D'Agostino and Stephens (1986, Chapter 2).

1.4 Examples

In this section, we will give numerical examples demonstrating the use of some of the goodness of fit tests we have briefly discussed. Some of these will be considered again in more detail later, where more powerful tests will be applied.

Example 1.4.1. Weldon's Dice Data

This data set was discussed by Pearson in his classic 1900 paper. The data are reproduced in Table 1.1, and give the number of occurrences of a 5 or a 6 in 26,306 throws of 12 dice. We will return to this example again in Example 8.3.1.

If it is assumed that the dice are fair, then the null hypothesis H_0 is that the probability of a 5 or 6 in one throw is 1/3, and so the probability of r occurrences of a 5 or 6 in 12 throws is given by the binomial probability, $p_r = {}^{12}C_r(1/3)^r(2/3)^{12-r}$. From this probability the expected frequencies of r occurrences of a 5 or 6 in the 26,306 throws of 12 dice can be calculated. To check whether the deviations between observed and expected frequencies are more than could be expected by chance, Pearson's X^2 statistic X_P^2 can be calculated. A visual comparison is given in Figure 1.1.

Pearson (1900) obtained $X_P^2 = 43.9$ while our calculations give $X_P^2 = 41.3$. The difference is almost certainly due to Pearson having used expected frequencies rounded to the nearest integer. In either case use of the χ_{12}^2 distribution to obtain P values indicates substantial deviations between observed and expected frequencies. With such a large number of throws of

Table 1.1 Weldon's data

Number of dice in cast with 5 or 6 parts	Observed frequency
0	185
1	1149
2	3265
3	5475
4	6114
5	5194
6	3067
7	1331
8	403
9	105
10	14
11	4
12	0

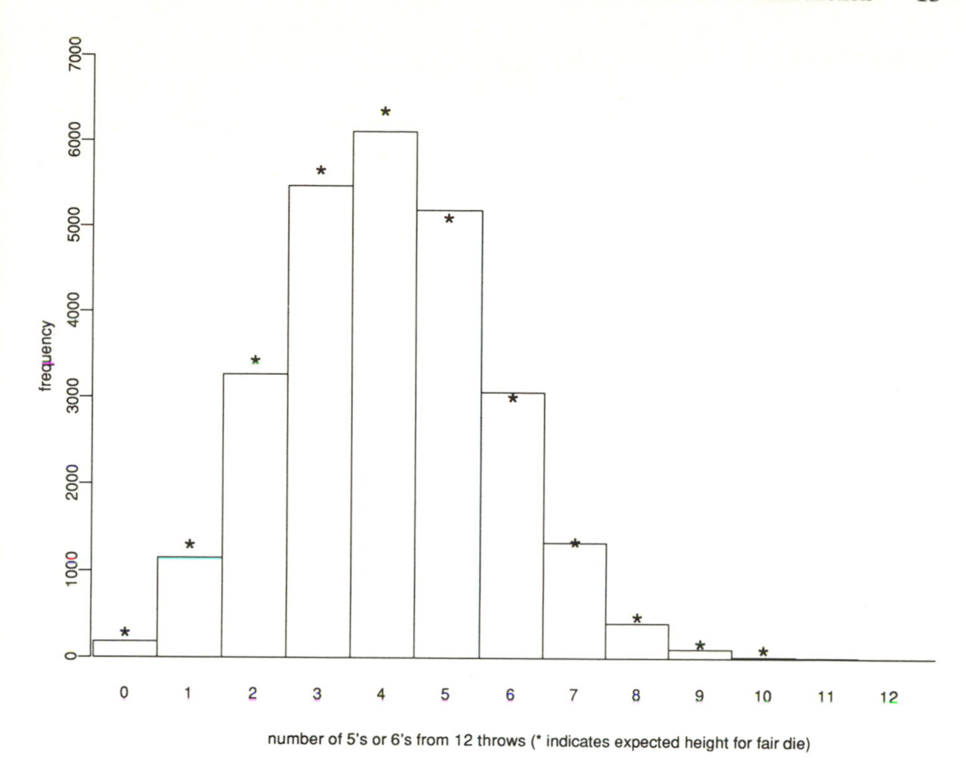

Figure 1.1 Histogram of Weldon's dice data.

12 dice this is hardly surprising. Closer inspection of Figure 1.1 indicates that there is a well defined trend for 5's or 6's to occur more than they should; it appears the dice were slightly biased.

Like most, if not all of the data sets to which Karl Pearson applied his X^2 test, this is a large data set. As Figure 1.1 shows, there is a definite trend or smooth alternative in this data. This may not have been picked up in a similar but smaller data set. Our next example will further highlight this. In Chapters 5 and 7 we will illustrate how components of X_P^2 complement the inspection of Figure 1.1 and spell out what the alternative is.

Example 1.4.2. Birth-Time Data

Unlike the previous example, data are often available ungrouped. In such cases application of Pearson's X^2 test is dubious because the data must be grouped and this loses information. Moreover, there is the problem of how to construct the groups. Suppose we consider a simple case when no estimation is required and ask whether birth-times occur uniformly throughout the day. Hospital administrators would be interested in the answer to this question. Mood et al. (1974, p. 509) gave the following times for 37

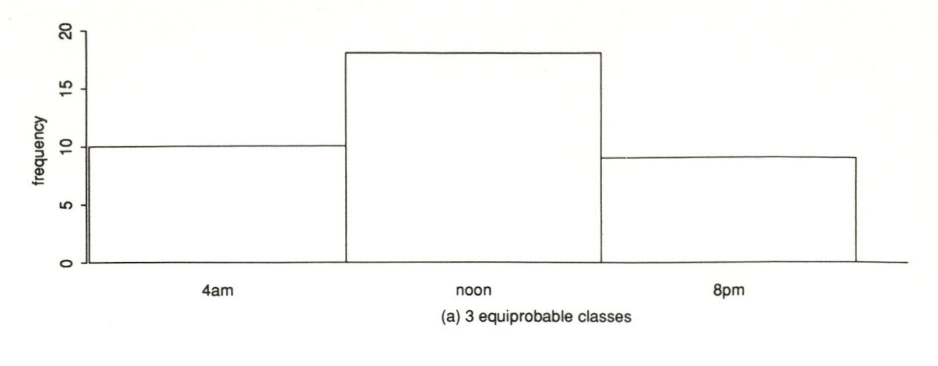

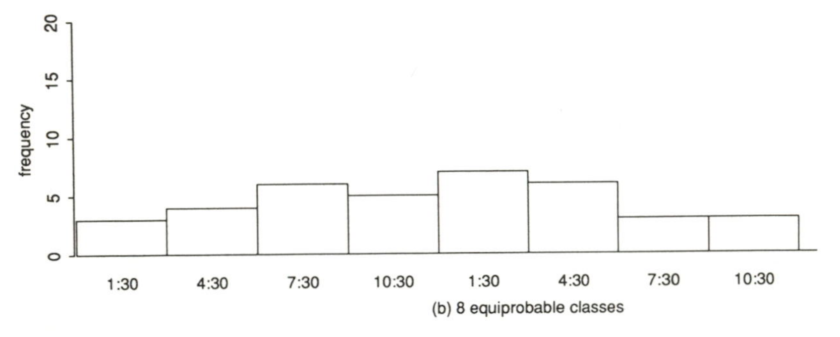

Figure 1.2 Histograms of birth-time data.

consecutive births:

> 7.02 P.M., 11.08 P.M., 3.56 A.M., 8.12 P.M., 8.40 A.M., 12.25 P.M., 1.24 A.M., 8.25 A.M., 2.02 P.M., 11.46 P.M., 10.07 A.M., 1.53 P.M., 6.45 P.M., 9.06 A.M., 3.57 P.M., 7.40 A.M., 3.02 A.M., 10.45 A.M., 3.06 P.M., 6.26 A.M., 4.44 P.M., 12.26 A.M., 2.17 P.M., 11.45 P.M., 5.08 A.M., 5.49 A.M., 6.32 A.M., 12.40 P.M., 1.30 P.M., 12.55 P.M., 3.22 P.M., 4.09 P.M., 7.46 P.M., 2.28 A.M., 10.06 A.M., 11.19 A.M., 4.31 P.M.

Figure 1.2 gives histograms based on three and eight equiprobable classes for this small data set. Both histograms indicate a trend toward more births during the day. With such a small sample size, however, "by eye" inspection can easily lead to false conclusions. In fact $X_P^2 = 3.95$ for three classes and a smaller value for eight classes; neither of these are significant although the X_P^2 value for three classes gives a much smaller P value than that for eight classes. Table 1.2 gives X_P^2 values for $p_r = 1/k$, $r = 1, \ldots, k$, when $k = 2$, 3, 4, 6 and 8, along with exact and approximate P values. Exact calculations for P values may differ from those based on the χ^2 approximation for small samples such as this, although, as can be seen from Table 1.2, the approximation is often not too bad, particularly with equiprobable p_r and use of a continuity correction. Details of the exact calculations are given in Best and Rayner (1982). Notice the variation in P values with k and that the k/n correction improves the crude χ^2 approximation to the P values.

Table 1.2 Birth data evaluaton of X_P^2 and P value for various k and $p_r = k^{-1}$, $r = 1, \ldots, k$

k	X_P^2	$P(\chi_{k-1}^2 \geq X_P^2)$	$P(\chi_{k-1}^2 \geq X_P^2 - k/n)$	Exact P
2	0.03	0.87	1.00	1.00
3	3.95	0.14	0.15	0.16
4	3.54	0.32	0.34	0.34
6	5.00	0.42	0.45	0.44
8	3.86	0.80	0.84	—

Perhaps other choices of p and k would give smaller P values more in line with the visual inspection of the plots. This example points to a difficulty with X^2 tests applied to small samples: as mentioned in the previous example, they are not good at picking out trends. The same numerical value of X^2 would result if the histogram bars were permuted so that no trend towards more daytime births was evident. In fact, this is one of the deficiencies that motivated Neyman's (1937) smooth test.

Given that the data are ungrouped, it seems reasonable to expect that tests not based on groups would be better at detecting any deviation from the null hypothesis. Indeed, Mood et al. (1974, p. 510) applied a Kolmogorov–Smirnov test. Their test is also nonsignificant, although their cumulative density plot is again suggestive of a "bulge" during the day. They do not comment on this.

Neyman's test statistic, mentioned in §1.2, is applied to this data in §4.4.

Example 1.4.3. Chemical Concentration Data

Risebrough (1972) gave data on the concentrations of various chemicals in the yolk lipids of pelican eggs. For 65 Anacapa birds the concentrations of polychlorinated biphenyl (PCB), an industrial pollutant, were:

452, 184, 115, 315, 139, 177, 214, 356, 166, 246, 177, 289, 175, 296, 205, 324, 260, 188, 208, 109, 204, 89, 320, 256, 138, 198, 191, 193, 316, 122, 305, 203, 396, 250, 230, 214, 46, 256, 204, 150, 218, 261, 143, 229, 173, 132, 175, 236, 220, 212, 119, 144, 147, 171, 216, 232, 216, 164, 185, 216, 199, 236, 237, 206, 87.

Can these data be summarized by saying they are normally distributed with mean and variance taken from the sample data? Figure 1.3 gives an equal width (the usual) histogram, and the probability density function of the normal distribution with the same mean and variance as the data. The histogram is clearly asymmetric. Thomas and Pierce (1979) report for this data set that both of their smooth tests were highly significant. They do not discuss what sort of departures from normality are evident. Using the $\hat{S}_4$ test of Best and Rayner (1985b) for this data set, $\hat{S}_4 = 9.89$ with components $\hat{V}_3 = 2.31$, $\hat{V}_4 = 1.99$, $\hat{V}_5 = 0.41$ and $\hat{V}_6 = -0.65$. The distribution of $\hat{S}_4$ is

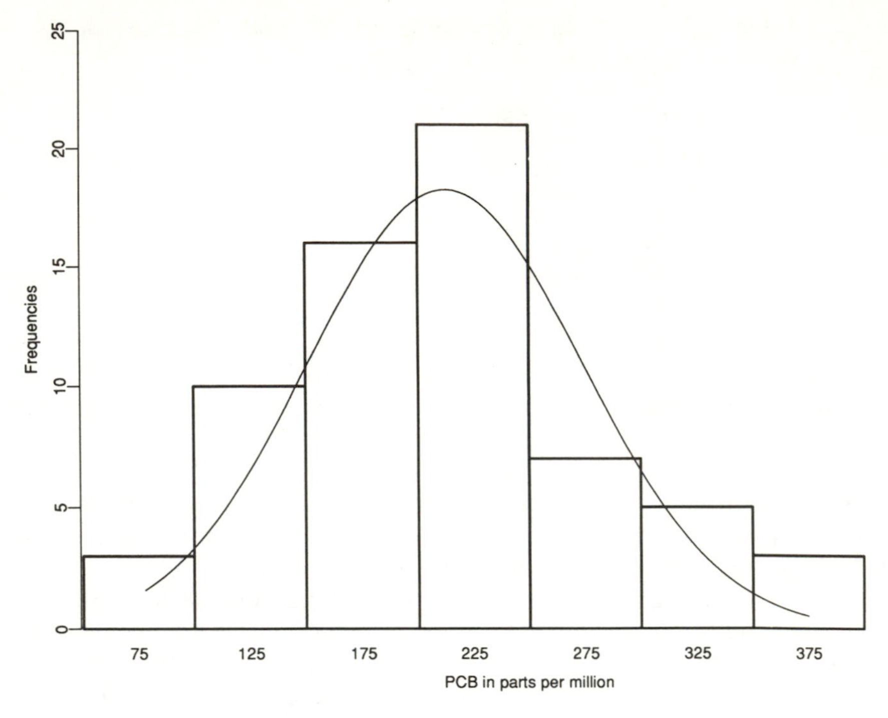

Figure 1.3 Histogram of PCB data.

approximately χ_4^2 and a P value of about .05 is indicated. Note that $\hat{V}_3 = \sqrt{(nb_1/6)}$ and $\hat{V}_4 = \sqrt{(n/24)}(b_2 - 3)$, where b_1 and b_2 are the usual measures of skewness and kurtosis. From the high $\hat{V}_3$ value we confirm the impression from Figure 1.3, which indicated skewness departures from normality.

Karl Pearson may not have applied his X^2 test to this data set as the sample size is small. But by the late 1920s R. A. Fisher had developed small sample goodness of fit tests based on moments that might give similar results to our $\hat{S}_4$ test. Fisher, however, did not consider optimality or the power of his tests.

Example 1.4.4. Mississippi River Data

The maximum daily rates of discharge of the Mississippi River measured at Vicksburg in units of cubic feet per second for the 50 years 1890–1939, after reordering from smallest to largest, were:

760, 866, 870, 912, 923, 945, 990, 994, 1018, 1021, 1043, 1057, 1060, 1073, 1185, 1190, 1194, 1212, 1230, 1260, 1285, 1305, 1332, 1342, 1353, 1357, **1457**, 1397, 1397, 1402, 1406, 1410, 1410, 1426, 1453, 1475, 1480, 1516, 1516, 1536, 1578, 1681, 1721, 1813, 1822, 1893, 1893, 2040, 2056, 2334.

The values, including the obvious misprint 1457, are taken from Gumbel

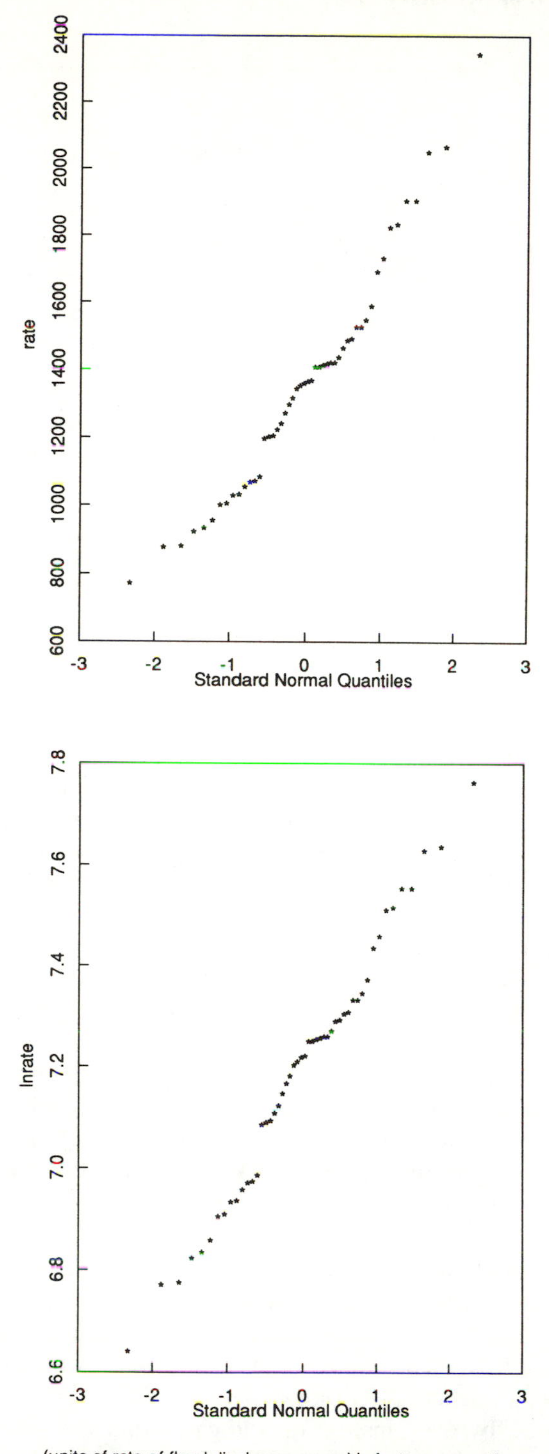

(units of rate of flood discharge are cubic feet per second)

Figure 1.4 Normal Q–Q Plot, Mississippi River data.

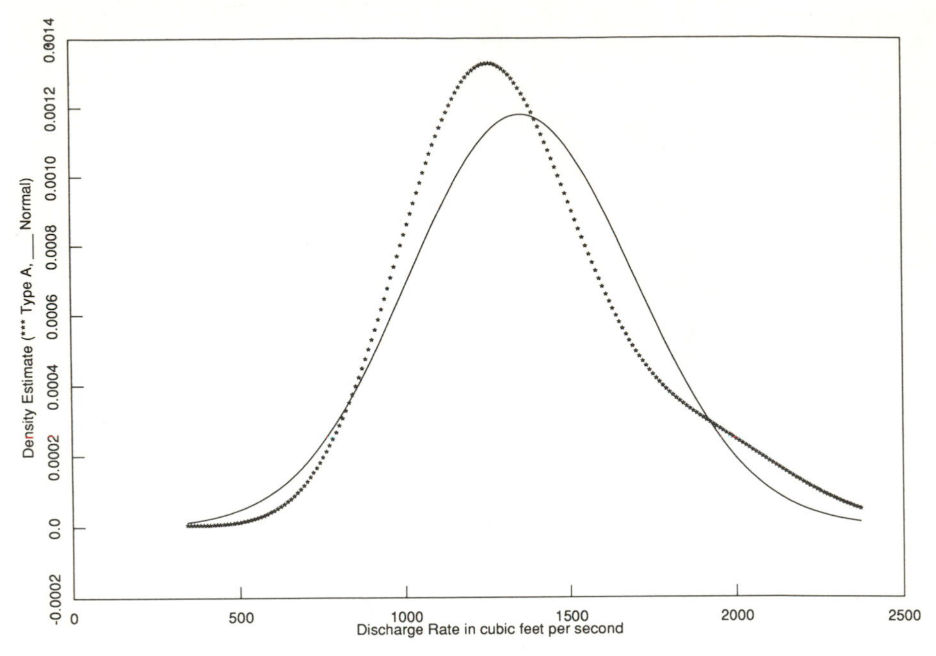

Figure 1.5 Density estimates for Mississippi data.

(1943). We assume the observations are independent. If a satisfactory probability model or distribution could be found for this data, then such a model would be useful for predicting the size of, say, a 1 in 100 year flood. Such predictions of course, have implications for life and property protection. With a sample size of only 50 a histogram is not always the best graphic technique for illustrating the data. Different class sizes and/or choice of number of classes can give different interpretations. Another commonly used technique is to plot the sorted data against quantiles of a standard probability distribution. Such plots are discussed in some detail by Wilk and Gnanadesikan (1968), Chambers et al. (1983, Chapter 6), and D'Agostino and Stephens (1986, Chapter 2). If the standard distribution is the normal then the plot is often called a *normal Q–Q plot.* The closer the points are to a straight line, the more likely are the data to be normal. The left half of Figure 1.4 gives a normal Q–Q plot for the Mississippi data. The curvature evident is a clear indication of the asymmetry of the data. This is confirmed by noting that although $\hat{S}_4$ is an insignificant 3.96, one of its components, the skewness component $\hat{V}_3$, accounts for most of its value. In fact $\hat{V}_3^2/\hat{S}_4 = 0.82$. Previous studies have indicated that by taking logarithms to the base *e* the data may produce a more normal distribution. The right of Figure 1.4 gives a normal Q–Q plot of the logarithms of the Mississippi data {ln(rate)} and a "by eye" inspection indicates that the plot is more linear. Calculation of $\hat{S}_4$ and its components confirms that the logarithms may be considered to be normally distributed.

Figure 1.5 shows the probability density function for a normal random variable with the same mean and variance as the data, and also the Gram–Charlier Type A series estimate of the density for the data. Gram–Charlier Type A series estimates are discussed in §8.2. The fitted density is more skewed than a normal density, confirming our previous analysis and the plots in Figure 1.4.

We reiterate that histograms, Q–Q plots, density estimates and other graphic techniques are useful adjuncts to significance tests in assessing the adequacy of models.

2
Pearson's X^2 Test

2.1 Introduction

Pearson's X^2 test was the first goodness of fit test. Moreover, it was one of the first tests of statistical inference, and perhaps is one of the most frequently used statistical tests. We therefore begin this chapter with a review of Pearson (1900), the paper in which Pearson announced the discovery of the test that is, according to Cochran (1952), "... one of the foundations of modern statistics." This is followed by a review of the properties of the test and developments that make it suitable for testing composite hypotheses. Further examples are then given of the application of these tests.

This chapter is intended as a review of a considerable volume of material, and no effort is made to *prove* results. Several of the tests may be considered as smooth tests, but their presentation as such will be left for later chapters (for example, see §§4.2 and 5.1).

2.2 Foundations

Before Karl Pearson introduced the X^2 goodness of fit test (Pearson, 1900), model assessment was a chancy business. Pearson himself, in Pearson (1897), could only conclude:

> Considering the rough nature of cloudiness observations, the agreement must be considered fairly good, and very probably the smooth results of the theory are closer to the real facts of the case than the irregular observations. (See Plackett: 1983, p. 61.)

The "considering" here is purely subjective. Three years later Pearson critized this sort of subjectivity:

> But the comparison of observation and theory in general amounts to a remark—based on no quantitative criterion—of how well theory and practice really do fit! (See Pearson, 1900, p. 171.)

The need for objectivity in model assessment is no less today than it was a century ago, but thankfully we now have the tools. That we do have the tools is very much due to Karl Pearson.

Before we look more carefully at this paper, it is as well to put Pearson's achievement in context. This is done in detail in Plackett (1983) who, in part, said

> During the period 1893–9, Pearson's statistical interests expanded rapidly. He developed his system of frequency curves, laid the foundations of multiple correlation and regression, and obtained the probable errors and correlations of estimated coefficients in large samples. Everything was directed towards the study of problems in heredity and evolution. Among these manifold activities is included his presentation of the multivariate normal distribution. (p. 61)

Presumably the "problems in hereditary and evolution" provided the motivation for the X^2 test and, as we shall see, much of the remainder of the work in this period was relevant in its derivation.

Pearson (1900) started somewhat abruptly into a "preliminary proposition," now sometimes called *Pearson's lemma.* This states that a particular quadratic form in multivariate normal random variables has a χ^2 distribution. A more modern discussion is given in Kendall and Stuart (1977, §15.10).

In the second section of the paper, integration by parts was used to develop a power series expansion for the χ^2 tail probabilities.

Probabilities were given to six decimal places, for degrees of freedom from 2 to 19 and gave "the probability of a system of deviations as great or greater than the outlier in question."

The X^2 test was derived in the third section. Observations were grouped into $n + 1$ categories, so that there were n algebraically independent random variables. The multinomial variances and covariances were cited, and then a polar transformation used to invert the covariance matrix to apply the "preliminary proposition" (Pearson's lemma). Of course, this assumed the multinomial approaches the multivariate normal, and that the sample size is sufficiently large. The quadratic form turned out to be

$$X_P^2 = \sum \text{(observed} - \text{expected)}^2/\text{expected}$$

which we briefly discussed in §1.2. Alternative derivations have been given in Kendall and Stuart (1977, example 15.3) and Lancaster (1969, §V.3)

Pearson now concluded that this "result is of very great simplicity, and very easily applicable." He then outlined, in three stages, how to apply the result.

Then, in §5, the vexed question of estimating parameters was addressed.

Unfortunately, Pearson concluded that estimation makes no difference, and this question was only properly resolved by Fisher in a series of papers in the 1920s.

Next followed two sections of illustrations. The first contained three illustrations where estimation was not required, while the other had five illustrations where estimation was required. The paper finished with a brief conclusion. Of the illustrations, it is informative to quote Cochran (1952, p. 319).

> Finally, the paper contains eight numerical applications of the new technique. In two of these he pokes fun at Sir George Airy and Professor Merriman. They had both published series of observations which they claimed to be good illustrations of variates that follow the normal distribution. In the absence of any test of goodness of fit, Airy and Merriman could judge this question only by eye inspection. Pearson showed that the significance probability for Airy's data was 0.014, although the data from which Pearson calculated X^2 had already been smoothed by Airy. Merriman fared worse, his probability being $1\frac{1}{2}$ parts in a million. These examples show the weak position in which the scientist was placed when he had to judge goodness or badness of fit in the absence of an objective test of significance.

In Pearson (1901), Pearson admitted to a numerical error in assessing Merriman's data. Apparently the odds against Merriman's data being normal are 3,667 to 1 if 12 classes are used, and 6,915 to 1 if 11 classes are used. Pearson felt his criticisms of Merriman's work were still quite justified in spite of the changed probability. The data are given and discussed in Example 2.5.1.

For more detail on Pearson, the period, and the early development of the X^2 test, the reader is referred especially to Lancaster (1969, Chapters 1 and 5) and Plackett (1983). Our intention here was to acquaint the reader with the roots of the subject. Although our methods are quite naturally, very different from those of Pearson, the achievements of that original contribution are great. Even today, Pearson's X^2 test could be claimed to be one of the most widely known tests of statistical inference. As Plackett (1983) concluded

> Pearson's 1900 paper on chi squared is one of the great monuments of twentieth-century statistics. (p. 70)

2.3 The Pearson X^2 Test—An Update

In this section modern developments in the theory and properties of Pearson's X^2 test are reviewed. This is done in two parts. The first, which concentrates on the notation, definition of the test, and the number and construction of classes for Pearson's test, should be read by all readers. The second may be omitted on their first reading by those with less interest in the theory. It develops matters related to the power of the test.

We preface this discussion by pointing out that general review material on Pearson's X^2 test is contained in Cochran (1952, 1954), Lancaster (1969), Horn (1977), and Hutchinson (1979). We will concentrate on more recent developments in this chapter.

2.3.1. Notation, Definition of the Test, and Class Construction

In the Pearson X^2 test for a simple null hypothesis it is desired to test the null hypothesis that a random sample $X_1, \ldots, X_n$ of size n comes from a population with completely specified cumulative distribution function $F(x)$, against a general alternative, not $F(x)$. The sample space is partitioned into m classes, and N_j is defined as the number of observations from the sample that fall into the jth class. If p_j is the probability of falling into the jth class when $F(x)$ holds, the Pearson X^2 test statistic is defined as

$$X_P^2 = \sum_{j=1}^{m} (N_j - np_j)^2/(np_j)$$

As n increases the distribution of X_P^2 tends to be increasingly well approximated by χ^2_{m-1}, the χ^2 distribution with $m-1$ degrees of freedom. Thus, if $X_P^2 > a_{m-1}(\alpha)$, the $100\alpha\%$ point of χ^2_{m-1}, then the null hypothesis that we are sampling from $F(x)$ is rejected at the $100\alpha\%$ level.

If the distribution specified by the null hypothesis is continuous, then a goodness of fit test appropriate for continuous data could be used, as there is presumably some loss of power in ignoring some of the information and categorizing the data. We will discuss several tests appropriate for continuous data in later chapters. Nevertheless, some users will prefer to use an X^2 test under these circumstances, perhaps because X^2 tests are familiar, available in a computer package, or convenient in that it may be easy to calculate the test statistic and that this test statistic has a conveniently tabulated null distribution.

If an X^2 test is to be used, then the cells or classes must be constructed. For example, as Kempthorne (1967) has pointed out, different conclusions may be reached if different constructions are used. One of the first constructions recommended that the expected cell frequencies, np_j, should all be at least 5 (for example, see Fisher, 1925, p. 84). If this is so, the asymptotic null distribution of X_P^2, namely χ^2_{m-1}, is a reasonable approximation for small n. Recommendations based on the criterion of accurately approximating the null distribution of X_P^2 have appeared regularly over the years: see the references in Holtzman and Good (1986), and especially Roscoe and Byars (1971), Larntz (1978), Koehler and Larntz (1980), and Lawal (1980). The Roscoe and Byars (1971) recommendations are:

1. With equiprobable cells, the average expected cell frequency should be at least 1 (that is, $n \geq m$) with a 5% significance level, and should be at least 2 (that is, $n \geq 2m$) with a 1% significance level.

2. When cells are not approximately equiprobable, the expected cell frequencies in (1) should be doubled.
3. For $m = 2$ the X_P^2 test should be replaced by the test based on the exact binomial distribution.

In general, we suggest that the correct way to evaluate the χ^2 approximation is to use what Radlow and Alf (1975) called the *exact* χ^2 *test* method. This assigns a P value to the probability of rejecting all X^2 values greater than or equal to the observed. This is *not* the method of evaluating the approximation that is used by a number of authors, including Neyman and Pearson (1931) and, more recently, Tate and Hyer (1973).

For the equiprobable Pearson X^2 test, with $p_j = m^{-1}$, Katti (1973) and Smith et al. (1979) have tabulated the exact null distribution of this test for $n \leq 50$ and various m. At the "usual" significance levels the nominal and exact sizes are usually acceptably close, but using these tables the exact size can be given. Holtzman and Good (1986) recommended the adjusted χ^2 approximation in the equiprobable case: namely if y_0 is an observable value of X^2 then

$$P(X_P^2 \geq y_0) \approx P(Y \geq y_0 - m/n \mid Y \text{ is distributed as } \chi^2_{m-1})$$

Until recently, Lancaster (1980) was virtually alone in not recommending the equiprobable test, if Pearson's test is to be used. He suggested that the interval over which the alternative is defined should be subpartitioned where the slope of the alternative probability density function is largest, and should not be partitioned at all if the slope is less than unity. More recently, Kallenberg (1985) also recommended nonequiprobable X^2 tests (see the discussion later in this section).

For the equiprobable Pearson's X^2 test, the problem is how to choose m. Mann and Wald (1942) derived the formula

$$m = 4\{2(n-1)^2/c^2\}^{(1/5)}$$

where the test is performed at the $100\alpha\%$ level of significance and c is the $100\alpha\%$ point of the standard normal distribution. Several subsequent authors, including Williams (1950) and Schorr (1974), have recommended a much smaller m than Mann and Wald suggested (see also Harrison, 1985.) Dahiya and Gurland (1973) observed that the best m depends on the alternatives. The authors have found that for a sequence of fixed-level Pearson X^2 tests with $m = 2, 3, 4, \ldots,$ the power typically increases to a maximum at, say, m_0 then decreases. This m_0 is frequently quite small, at about 4 or 5. Using 20 plus classes, as the Mann–Wald formula often suggests, will then cause a substantial power loss. As a counter to these negative points, we note that the window width formula used in kernel density estimation as outlined in Silverman (1986, chapter 3) also involve a 1/5 power.

2.3.2. Power Related Properties

If a particular alternative to the null hypothesis holds, and gives π_j as the probability of falling into the jth class, then the asymptotic nonnull distribution of X_P^2 is $\chi^2_{m-1}(n\phi^2)$, noncentral χ^2 with $m-1$ degrees of freedom and parameter of noncentrality $n\phi^2$, where

$$\phi^2 = \sum_{j=1}^{m} (\pi_j - p_j)^2/p_j$$

This distribution may be used to calculate approximate powers, but is not as accurate as a simple three moment χ^2 approximation investigated by Best, Rayner, and Turnbull (1983). For small n they recommended either exact calculation or simulation. For larger n, approximate $c_0 + c_1 X_P^2$ by χ^2_ν, where, by equating corresponding moments,

$$c_1 = 4\mu_2/\mu_3, \qquad \nu = c_1^2\mu_2/2 \quad \text{and} \quad c_0 = \nu - c_1\mu$$

Here μ is the mean and μ_2 and μ_3 are the second and third central moments of X_P^2 under the alternative hypothesis, and are given by Koehler (1979).

However, tests are usually preferred only if they have "good" power or efficiency. The desirable power properties of the equiprobable Pearson test include that it is strictly unbiased and type D (a local optimality condition) as a multinomial test (for details, see Cohen and Sackrowitz, 1975). However it is not type D when used as a test of goodness of fit, as Rayner and Best (1982) showed. It is appropriate in a distance sense (see Spruill, 1976) and if the entropy is to be minimized (Cox and Hinkley, 1974). We later show that the Pearson X^2 test is a score test, and is thus asymptotically optimal for the model adopted.

Power simulations in Kallenberg (1985) and elsewhere supported Lancaster, in that appreciable power gains were achieved using nonequiprobable tests for heavy tailed alternatives; however, we have two reservations. If the alternative were known, a more refined analysis, not using X_P^2 would probably be used. Also, as we have reported, using more classes sometimes reduces the power. For a given alternative, therefore, six classes with boundaries at points of steep slope of the alternative probability density function may be more powerful than six equiprobable classes. But ten classes, no matter how they are chosen, may well be less powerful. The classes should be formed to be *tail discriminating*—that is, there should be relatively more, and hence lower probability classes, in the heavy tails.

The work of Kallenberg et al. (1985), Kallenberg (1985), and Oosterhoff (1985) attempted to explain the mechanisms for this behavior. Using local and nonlocal asymptotic theory on contamination and exponential families, it was shown that for heavy-tailed alternatives, a larger m produced greater asymptotic power, while for lighter-tailed alternatives a smaller m produced greater asymptotic power. The asymptotic models, however, do not necessarily agree with finite sample calculations. The statistician is typically presented with a sample of fixed size, and must choose m. This needs to be reconciled with asymptotics in which m is fixed and $n \to \infty$, or both m and $n \to \infty$ in some functionally dependent way. Our approach is to look at the basis of the parameter space for fixed n and m (see Chapter 3 and §§4.3 and 5.2). Nevertheless, choosing a large m for heavy-tailed alternatives does agree with the limited simulation studies we have seen.

For Pearson X^2 tests we have suggested, in Best and Rayner (1981, 1982, 1985a), various alternatives as to how to choose m. These are:

1. To perform a sequence of equiprobable X^2 tests with an increasing number of classes.
2. Use the *components* of X_P^2, checking for residual variation. These components are defined in Chapter 5.

As an example of a situation where (1) is applicable, consider a preliminary investigation when testing a new random number generator. The sequence of X^2 tests will determine a number of classes m_0 that is most critical of the data. Effectively, a class of alternatives is being singled out—the alternatives one hopes to most powerfully detect in the subsequent investigation. An example of option (2) is given in Best and Rayner (1985b). A typical outcome would be that particular components are significantly large. In subsequent investigations, these can then be focused upon to more powerfully detect particular alternatives. Using components with as many classes as possible eliminates loss of information due to categorization. On the other hand, option (1) is easier and more familiar to many users.

Whether the data are categorized or not, the statistician has to choose the number and width of the classes. The preceding remarks should not be taken to imply that we recommend the use of X^2 tests for continuous data. On the contrary, we recommend smooth tests of the form to be discussed in Chapters 4 and 6. We *are* pointing out that in selecting the classes to be used for an X^2 test, the user is selecting a class of alternatives that the chosen test best detects.

The Pearson X^2 test is a member of the power divergence family of statistics introduced in Cressie and Read (1984). They defined

$$2n\, I^{\lambda}(X \mid n : p) = 2/\{\lambda(\lambda+1)\} \sum_{j=1}^{m} X_j\{[X_j/(np_j)]^{\lambda} - 1]\}$$

For $\lambda = 1, 0, -0.5$ this produces X_P^2, the likelihood-ratio test statistic, and the Freeman–Tukey statistic, respectively. The performance of this class is further examined by Read (1984a, 1984b). These studies give perspective to the properties of Pearson's test. From the point of view of Bahadur efficiency, the likelihood-ratio test is the preferred member of the family; by Pitman efficiency, the Pearson test is best. Comparisons between these two protagonists abound in the literature; for example, see West and Kempthorne (1971), Kallenberg et al. (1985), Kallenberg (1985), and Oosterhoff (1985). Rather than delve into these riches, we note that Moore (1986) recommended the Pearson test of those in the power divergence family. In later chapters we hope to add to the desirable properties of the Pearson test, and so add weight to that recommendation.

Finally we note the work of Gleser and Moore (1985), who showed that failure of the observations to be independent results in incorrect significance levels. In particular, in Gleser and Moore (1985, p. 460) they said,

> positive dependence among successive observations causes all Pearson-type tests for categorical data to reject a null hypothesis too often.

2.4 X^2 Tests of Composite Hypotheses

Suppose that the null distribution depends on a vector $\beta = (\beta_1, \ldots, \beta_q)^T$ of unknown or "nuisance" parameters. Then in X_P^2, p_i must be replaced by an estimate of p_i, say $\hat{p}_i$. If the $\hat{p}_i$ are based on the grouped maximum likelihood estimators, say β^*, we have $\hat{p}_i = p_i(\beta^*)$. The new statistic is X_{PF}^2, the Pearson–Fisher statistic, which has an asymptotic null χ^2_{m-q-1} distribution and an asymptotic $\chi^2_{m-q-1}(n\hat{\phi}^2)$ alternative distribution, where $\hat{\phi}^2$ is simply ϕ^2 with both π_j and p_j replaced by cell probabilities using the estimate β^* of the nuisance parameter β. Although X_{PF}^2 *may* be less powerful than its competitors (see Rao and Robson 1974, but note the discussion following in this section and in §7.4), it would seem to provide a more robust test. This is because X_{PF}^2 depends on the null hypothesis only in the placement of the cell boundaries and through the estimates, but not in its functional form.

If maximum likelihood estimators based on the ungrouped observations are used, we obtain the Chernoff–Lehmann (1954) statistic X_{CL}^2. The null distribution is χ^2_{m-q-1} plus a linear combination of χ^2_1 variables, and depends on the unknown nuisance parameters. It is sometimes sufficient to base inferences on the fact that the null distribution of X_{CL}^2 is bounded between χ^2_{m-q-1} and χ^2_{m-1}. If it is not sufficient, then different tables of null percentage points are required for each family of distributions for which we test. Using random cell boundaries in X_{CL}^2 in the manner described in Roy (1956) and Watson (1959) yields the X_R^2 statistic. In X_R^2 the dependence of the null distribution on the nuisance parameters is removed, but not the need for tabulation of critical points for each family tested. Dahiya and Gurland (1972) treated the normal case.

Finally, X_{RR}^2 is the Rao–Robson statistic, introduced by Rao and Robson (1974). The form of X_{RR}^2 will be given in §6.7, where it will be apparent that it depends on the null probability density function; it is, in fact, a score statistic. X_{RR}^2 may be viewed as X_{CL}^2 plus a correction factor to recover the lost degrees of freedom. Its null distribution is χ^2_{m-1} no matter how many parameters are estimated.

Moore (1977) gave a "recipe" for producing goodness of fit statistics with χ^2 distributions. These may have fixed or random cells. X_{PF}^2 and X_{RR}^2 are both produced by this method.

It seems to be usually assumed, as in Kopecky and Pierce (1979), that the

X^2_{PF} test is less powerful than those based on X^2_{CL} and X^2_R, which are less powerful than the X^2_{RR} test. The evidence is a series of particular cases, and the point of Moore and Spruill (1975, p. 615), that the X^2_{PF} test is neither always better nor always worse than those based on X^2_{CL} and X^2_{RR}, is overlooked. It appears that for any particular alternative, various statistics have different degrees of freedom and different parameters of noncentrality, and the most powerful test is determined by how each factor compensates for the other. An explanation of the different power performances is given in Chapters 4 and 5, in terms of the bases of the parameter space.

Another route to these X^2 statistics is via quadratic score statistics. Kopecky and Pierce (1979, §3) showed that some of the Pearson-type statistics mentioned earlier are particular cases of quadratic score statistics. Thomas and Pierce (1979) pointed out that these statistics are generalized Neyman statistics with a "correction term" to adjust for estimation. If the nuisance parameters that enter the problem are location-scale parameters, then the suggested statistics do not involve the nuisance parameters. For each family tested, however, a different form of the test statistic must be calculated.

A more comprehensive account of the material in this section was given by Moore in chapters in Hogg (1978) and in d'Agostino and Stephens (1986). Also, see the latter for detail on various goodness of fit tests, such as those based on the empirical distribution function and those based on correlation and regression. Most of the competitors to the smooth tests we recommend in later chapters were discussed in d'Agostino and Stephens (1986), and it would be repetitious to include the same material here.

2.5 Examples

In this section, we give three numerical examples of some of the goodness of fit tests discussed in the first two chapters.

All three involve testing for normality, and in all three we wish to calculate X^2_{PF}. This requires the use of the maximum likelihood estimates for the grouped data. It is tempting to use estimators μ^* of μ and σ^{*2} of σ^2 given by

$$\mu^* = \sum_j f_j x_j / n \text{ and } \sigma^{*2} = \sum_j f_j (x_j - \mu^*)^2 / n$$

where the f_j are the observed class frequencies and the x_j are the class midpoints. The estimators μ^* and σ^{*2} are only approximations to the actual maximum likelihood estimators $\hat{\mu}$ and $\hat{\sigma}^2$. Their use will cause X^2_{PF} to not have *asymptotic* distribution χ^2_{m-q-1}, but will give a good approximation to X^2_{PF}.

Cramer (1946, p. 438) suggested an improved estimator for σ^2 in the case when there are equal class widths (h). This is obtained by subtracting the usual Shepherd's correction, $h^2/12$, from σ^{*2}. For larger data sets where h is not too large it appears Cramer's suggestion is an excellent one. It is fortunate that the maximum likilihood estimators can be approximated in this way because iterative methods are needed to calculate $\hat{\mu}$ and $\hat{\sigma}^2$, and the software is usually not conveniently available.

We reiterate that for "large" data sets with h not too "small", using X^2_{PF} with estimators μ^* and $\sigma^{*2} - h^2/12$ is a reasonable approximate procedure for testing normality; however, we have not investigated what "large" and "small" mean. Perhaps "large" is too big to allow an approximate P value to be found via Monte Carlo simulation.

Example 2.5.1. Merriman's Target Data

This large data set was discussed by Pearson (1900, 1901) and is presented in Table 2.1. A possible null hypothesis is that the data are normally distributed with both the mean (μ) and the variance (σ^2) unknown. As only

Table 2.1 Merriman's data

Belt	1	2	3	4	5	6	7	8	9	10	11	Total
Observed frequency	1	4	10	89	190	212	204	193	79	16	2	1000

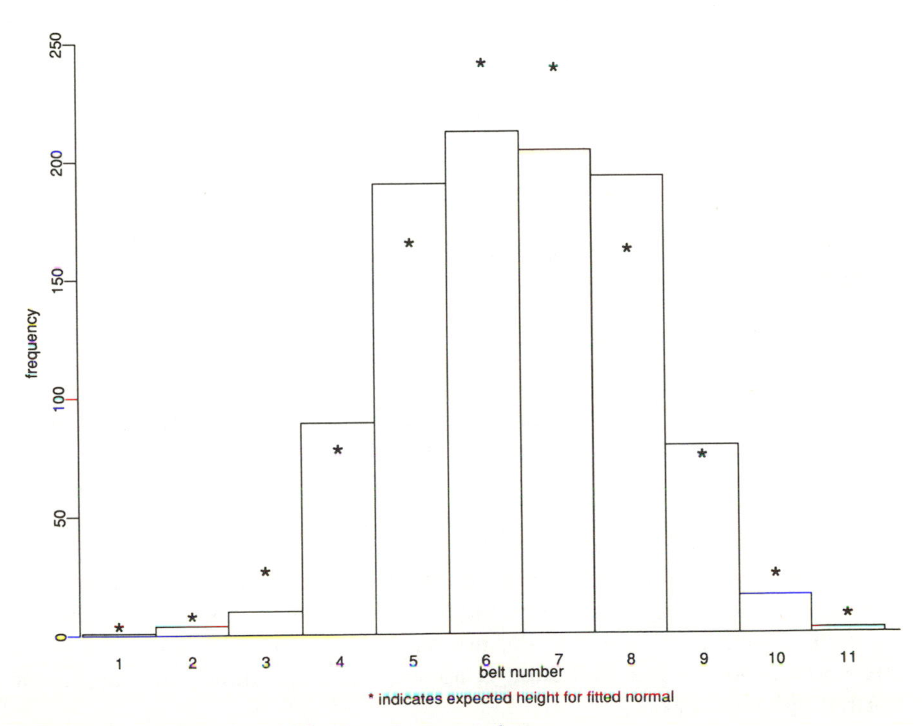

Figure 2.1 Histogram of Merriman's target data.

grouped frequency data are available the only test discussed in §2.4 which is applicable is that based on the Pearson–Fisher X^2_{PF} statistic. If class intervals of (0, 1], (1, 2], (2, 3], and so on are taken as applicable, then the estimators are $\mu^* = 5.982$, $\sigma^* = 1.577$, and $\sqrt{\{\sigma^{*2} - h^2/12\}} = 1.550$. Pearson (1900) obtained $X^2_{PF} = 45.8$ while our calculations give $X^2_{PF} = 34.36$ using σ^* and $X^2_{PF} = 34.64$ using $\sqrt{\{\sigma^{*2} - h^2/12\}}$. This is much closer to the corrected value of 34.63 given by Pearson (1901), but in either case the normality hypothesis cannot be maintained. Figure 2.1 gives a visual comparison that also suggests nonnormality of the observed data.

It is interesting to recall that the analysis of this data in Pearson (1900) is incorrect. As we stated in §2.2, Pearson used the wrong degrees of freedom for X^2_{PF}.

Example 2.5.2. Mothers' Heights Data

Pearson and Lee (1903) gave data on mothers' heights that is typical of much of the data used in the large heredity studies of the period. Snedecor and Cochran (1980, p. 50, Table 4.6.2) presented the frequency data. In a later example (Snedecor and Cochran 1980, p. 77, Example 5.12.5), they obtained an X^2 value of 11.85 based on 6 degrees of freedom. The question is, which X^2 statistic was used? If the ungrouped data were used to calculate $\hat{\mu}$ and $\hat{\sigma}$, as suggested in Snedecor and Cochran's Example 4.6.7, then the statistic should have been X^2_{CL}, and the null distribution of the test statistic

Table 2.2 Mothers' heights data

Class limits	Observed frequency	Class limits	Observed frequency
$(-\infty, 55]$	3.0	(63, 65]	277.5
(55, 57]	8.5	(65, 67]	119.5
(57, 59]	52.5	(67, 69]	23.5
(59, 61]	215.0	$(69, \infty)$	6.5
(61, 63]	346.0		

would *not* have been χ^2 with six degrees of freedom. If grouped data were used, then the statistic should have been X^2_{PF}, and the null distribution of the test statistic would have been χ^2 with six degrees of freedom. We calculate that $\mu^* = 62.486$, $\sigma^* = 2.434$, and $\sqrt{\{\sigma^{*2} - h^2/12\}} = 2.365$, giving $X^2_{PF} = 11.92$ using σ^* and $X^2_{PF} = 12.87$ using $\sqrt{\{\sigma^{*2} - h^2/12\}}$. This is in reasonable agreement with Snedecor and Cochran's figures, so it appears that they calculated X^2_{PF}. Our Figure 2.2 gives a histogram and the probability density function of the normal distribution with the same mean and variance as the grouped data. It appears there is some doubt about the normality hypothesis.

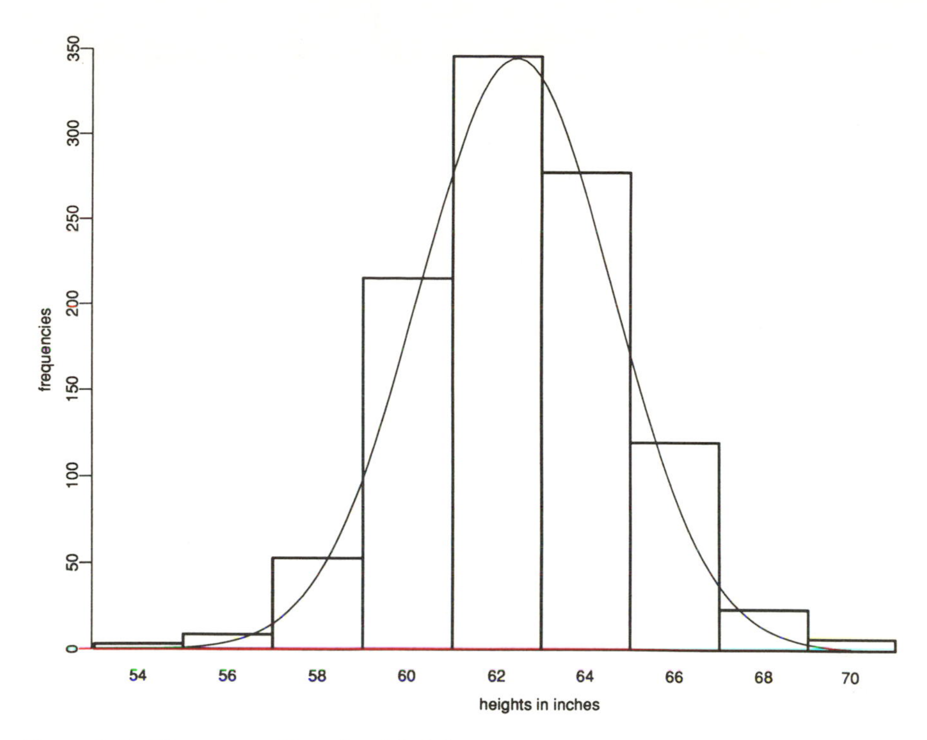

Figure 2.2 Histogram of mothers' heights.

Example 2.5.3. Chemical Concentration Data

This data was discussed in Example 1.4.3. The question is, Can the data be regarded as being normally distributed? Our previous conclusion was that the data were too highly skewed to retain the normality assumption.

All of the X^2 tests in §2.4 can be used to test normality, but we look at X^2_{PF} and X^2_R only, as we feel that not many applied statisticians would use X^2_{CL} or X^2_{RR}. To use X^2_{PF} we have to assume that classes had been chosen arbitrarily beforehand. Equal width classes are often chosen, and if we had $(50, 100]$, $(100, 150]$, ..., $(350, 400]$, with the proviso that the end classes contain any observations below 50 or greater than 400, then the observed frequencies would be 3, 10, 16, 21, 7, 5, and 3, as shown in Figure 1.3.

Estimating μ and σ^2 from these frequencies and class midpoints 75, 125, ..., 375 gives $\mu^* = 210.4$, $\sigma^* = 71.1$, and $\sqrt{\{\sigma^{*2} - h^2/12\}} = 69.6$. We obtain $X^2_{PF} = 4.34$ using σ^* and $X^2_{PF} = 4.57$ using $\sqrt{\{\sigma^{*2} - h^2/12\}}$. If the former is regarded as a χ^2_4 variate, we get an approximate P value of 0.36, which does not confirm our visual inspection of the asymmetric histogram in Figure 1.3, or the Gram–Charlier Type A density shown in Figure 2.3. As in §§1.4 and 8.2 the first four nonzero terms are used in this estimate. If we use X^2_R, however, we find from the ungrouped data that $\hat{\mu} = 210.14$ and $\hat{\sigma} = 72.9$. Thomas and Pierce (1979) reported that $X^2_R = 5.52$ and 7.77 for 8

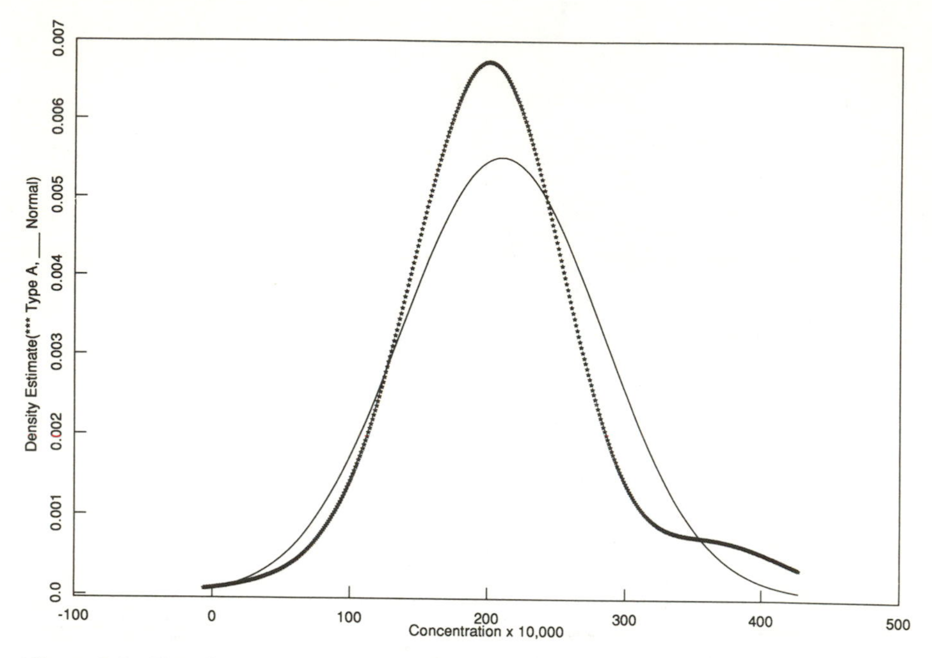

Figure 2.3 Density estimate for PCB data.

and 10 equiprobable classes, respectively, and again that $P > .10$ in both cases. These do not confirm the visual inspection of Figures 1.3 and 2.3 or the analysis of Example 1.4.3. Thomas and Pierce (1979) called X_R^2 the "ordinary χ^2 test," but perhaps X_{PF}^2 could still be given this title.

From our previous analysis, the smooth tests of Thomas and Pierce and $\hat{S}_4$ tests were better able to detect the nonnormality than the X^2 tests used here. That may be because the alternative is what we subsequently call "of low order." As we mentioned in Example 1.4.3, however, it is doubtful whether Karl Pearson would have used an X^2 test on such a small data set. Given the numerical analysis difficulties in finding $\hat{\mu}$ and $\hat{\sigma}$, plus the fact that we could have used an uncategorized test for this data, we also would not advise using X_{PF}^2 in this case.

3

Asymptotically Optimal Tests

3.1 Introduction

One of the criticisms aimed at X^2-type tests in the past has been that if the null hypothesis is rejected then there is no alternative distribution indicated. (For example, see Durbin and Knott, 1972, p. 291). Were this true, there would be no alternative model the user could adopt. In fact, this is not true for the Pearson X^2 test, nor for the tests we will consider in later chapters, where we will embed probability functions and probability density functions within so-called smooth families and derive tests that the null distribution adequately describes the data. If the null hypothesis is rejected then we may turn to the embedding family as an alternative model.

Since we have parametric models under both the null and alternative hypotheses, we could do as Neyman himself did in his 1937 paper introducing the smooth tests. He defined a sense of "optimal" suitable for any sample size, and derived the test that is optimal in this sense for the smooth model. We will give details of this approach in §4.1. In this chapter, we will outline the form of the likelihood ratio, score, and Wald tests. These tests have good properties in "large" samples (i.e., as the sample size becomes infinitely large). In subsequent chapters, one of these tests—the score test—is applied to deriving "optimal" smooth goodness of fit tests for both categorized and uncategorized parametric models, and also for cases where parameter estimation is and is not necessary.

The smooth tests we describe are in a sense optimal in large samples, and, when there are no nuisance parameters, optimal in a different sense for *any* sample size. This makes them more desirable than competitors that fail to satisfy these notions of optimality. We consider it unfortunate to see tests such as the Pearson X^2 test maligned because of "poor" power performances *when compared to tests suitable for continuous data* because the Pearson X^2 test is an optimal test of *categorized* data.

We will now briefly examine the large sample optimal tests—the likelihood ratio test, the Wald test, and the (quadratic) score test. These will be outlined for both simple and composite null hypotheses. Then, in the final section, the properties of these large sample tests and the sense in which they are optimal will be discussed. This is a more technical section and may be omitted at first reading.

In specific situations, it is as well to have the choice between the three large sample tests. Each requires that various estimators be calculated and some will be more convenient than others. In spite of the equivalence of the tests in large samples, their small sample properties will be different. In multivariate normal models, the likelihood ratio test is usually preferred; for the smooth models of this monograph, the score statistic is most convenient. In particular models, the statistician should be prepared to choose between the three large sample tests.

3.2 The Likelihood Ratio, Wald, and Score Tests for a Simple Null Hypothesis

An elementary exposition of these tests was given by Buse (1982); a not-so-elementary one was given by Cox and Hinkley (1974). Both here and in §3.3 we will aim for a level of treatment somewhere in between these.

To begin with, suppose we are given a random sample $X_1, \ldots, X_n$ from a continuous distribution with probability density function $f(x; \theta)$ in which $\theta = (\theta_1, \ldots, \theta_k)^T \in \Theta$, the parameter space. We aim to test the simple null hypothesis H_0: $\theta = \theta_0$ against the composite alternative K: $\theta \neq \theta_0$, where θ_0 is not on the boundary of Θ. If the distribution of X is discrete, then the following results are essentially the same. Note that realizations of random variables; such as X, are denoted by the corresponding lowercase letter, viz x. So the likelihood is the product of $f(x_1; \theta), \ldots, f(x_n; \theta)$; the corresponding random variable is $f(X_1, \theta) \times \ldots \times f(X_n; \theta)$.

The likelihood ratio test was proposed by Neyman and Pearson (1928), and is based on the statistic

$$L = 2l(\hat{\theta}; X) - 2l(\theta_0; X)$$

where $l(\theta; X)$ corresponds to the natural logarithm of the likelihood,

$$l(\theta; x) = \sum_{i=1}^{n} \log f(x_i; \theta)$$

and where $\hat{\theta}$ is the maximum likelihood estimator of θ, chosen to maximize the likelihood $\prod_{i=1}^{n} f(x_i; \theta)$ for all $\theta \in \Theta$. Now define

1. The *efficient score* $U(\theta) = (U_i(\theta))$, in which $U_i(\theta) = \partial l(\theta; X)/\partial \theta_i$
2. The *information matrix* $I(\theta) = (I_{ij}(\theta))$, in which

$$I_{ij}(\theta) = -E_\theta[\partial^2 l(\theta; X)/\partial \theta_i \, \partial \theta_j]$$

Note that expectation with respect to the probability density function

$f(x; \theta)$ when θ is arbitrary is denoted by E_θ; if θ is constrained to be 0 the expectation is denoted by E_0.

To test H_0 against K, Wald (1943) suggested the test statistic

$$W = (\hat{\theta} - \theta_0)^T I(\hat{\theta})(\hat{\theta} - \theta_0)$$

and Rao (1948) the test statistic

$$S = \{U(\theta_0)\}^T \{I(\theta_0)\}^{-1} \{U(\theta_0)\}$$

S does not require the calculation of the maximum likelihood estimator, but does require the existence of the inverse of the information matrix. The null hypothesis H_0 is rejected for large values of L, W, and S. Under the null hypothesis, all three statistics are asymptotically distributed as central χ_r^2 random variables, where r is the number of elements in θ, or, equivalently, the dimension of Θ.

Example 3.2.1.

Based on a random sample of size n from a Poisson distribution with parameter λ, we wish to test H_0; $\lambda = \lambda_0$ against $K: \lambda \neq \lambda_0$. We find the logarithm of the likelihood function is given by

$$l = -n\lambda + t \log \lambda - \sum \log(x_i!)$$

where t is a realization of $T = X_1 + \ldots + X_n$. Differentiation gives

$$\frac{\partial l}{\partial \lambda} = \frac{t}{\lambda} - n, \quad \text{and} \quad \frac{\partial^2 l}{\partial \lambda^2} = -\frac{t}{\lambda^2}, \quad \text{so that} \quad \hat{\lambda} = \frac{T}{n}$$

Then

$$U = T/\lambda - n \quad \text{and} \quad I = n/\lambda$$

from which

$$L = 2n(\lambda_0 - T/n) + 2T \log\{T/(n\lambda_0)\}$$

$$W = (T/n - \lambda_0)^2 n/\hat{\lambda} = (T - n\lambda_0)^2/T$$

$$S = (T/\lambda_0 - n)^2 \lambda_0/n = (T - n\lambda_0)^2/(n\lambda_0)$$

Clearly, the test based on S is equivalent to the test based on the normal approximation to T, which assumes T is normal with mean and variance both $n\lambda_0$ under H_0.

Example 3.2.2.

Suppose a random sample of size n is taken from a normal distribution with mean μ and variance σ^2, hereafter written $N(\mu, \sigma^2)$. We wish to test for a specific mean and specific standard deviation. Put $\theta = (\mu, \sigma)^T$, so that H_0 specifies $\theta_0 = (\mu_0, \sigma_0)^T$. The logarithm of the likelihood is

$$l = -\frac{n}{2} \log(2\pi) - n \log \sigma - \frac{\sum (x_j - \mu)^2}{2\sigma^2}$$

The following derivatives are obtained routinely:

$$\frac{\partial l}{\partial \mu} = \frac{\sum (x_j - \mu)}{\sigma^2} = \frac{n(\bar{x} - \mu)}{\sigma^2}, \qquad \frac{\partial l}{\partial \sigma} = -\frac{n}{\sigma} + \frac{\sum (x_j - \mu)^2}{\sigma^3}$$

$$\frac{\partial^2 l}{\partial \mu^2} = -\frac{n}{\sigma^2}, \qquad \frac{\partial^2 l}{\partial \mu\, \partial \sigma} = -\frac{2n(\bar{x} - \mu)}{\sigma^3}, \qquad \frac{\partial^2 l}{\partial \sigma^2} = \frac{n}{\sigma^2} - \frac{3\sum (x_j - \mu)^2}{\sigma^4}$$

The unrestricted maximum likelihood estimators are

$$\hat{\mu} = \bar{X} \text{ and } \hat{\sigma} = \sqrt{\left\{\sum (X_j - \bar{X})^2/n\right\}}$$

Thus the efficient score and information matrix are

$$U(\theta) = \begin{pmatrix} n(\bar{X} - \mu)/\sigma^2 \\ -n/\sigma + \sum (X_j - \mu)^2/\sigma^3 \end{pmatrix} \quad \text{and} \quad I(\theta) = \begin{pmatrix} n/\sigma^2 & 0 \\ 0 & 2n/\sigma^2 \end{pmatrix}$$

respectively. If we write $\tilde{\sigma} = \sqrt{\{\sum (X_j - \mu_0)^2/n\}}$, we may now calculate

$$L = 2n \log(\sigma_0/\hat{\sigma}) + n(\tilde{\sigma}^2/\sigma_0^2 - 1)$$

$$W = (\hat{\mu} - \mu_0, \hat{\sigma} - \sigma_0)\begin{pmatrix} n/\hat{\sigma}^2 & 0 \\ 0 & 2n/\hat{\sigma}^2 \end{pmatrix}\begin{pmatrix} \hat{\mu} - \mu_0 \\ \hat{\sigma} - \sigma_0 \end{pmatrix}$$

$$= n(\bar{X} - \mu_0)^2/\hat{\sigma}^2 + 2n(\hat{\sigma} - \sigma_0)^2/\hat{\sigma}^2$$

$$S = \left(\frac{n(\hat{\mu} - \mu_0)}{\sigma_0^2}, \frac{n(\tilde{\sigma}^2 - \sigma_0^2)}{\sigma_0^3}\right)\begin{pmatrix} \sigma_0^2/n & 0 \\ 0 & \sigma_0^2/(2n) \end{pmatrix}\begin{pmatrix} n(\hat{\mu} - \mu_0)/\sigma_0^2 \\ n(\tilde{\sigma}^2 - \sigma_0^2)/\sigma_0^3 \end{pmatrix}$$

$$= n(\bar{X} - \mu_0)^2/\sigma_0^2 + n(\tilde{\sigma}^2 - \sigma_0^2)^2/(2\sigma_0^4)$$

We can easily confirm the asymptotic distribution of W. For $\bar{X}$ is distributed as $N(\mu_0, \sigma_0^2/n)$ under H_0, so that $n(\bar{X} - \mu_0)^2/\sigma_0^2$ is χ_1^2 and $n(\bar{X} - \mu_0)^2/\hat{\sigma}^2$ is approximately χ_1^2. By maximum likelihood theory, $\hat{\sigma}$ is asymptotically distributed as $N(\sigma_0, \sigma_0^2/(2n))$ under H_0, so $2n(\hat{\sigma} - \sigma_0)^2/\sigma_0^2$ is also asymptotically distributed as χ_1^2, as is $2n(\hat{\sigma} - \sigma_0)^2/\hat{\sigma}^2$. The statistical independence of $\bar{X}$ and S^2 implies that of $\bar{X}$ and $\hat{\sigma}$, so W has an asymptotic χ_2^2 distribution.

Many other ways of combining these asymptotically independent χ_1^2 variates are possible, and these correspond to other statistics asymptotically equivalent to W and S, such as

$$W_0 = (\hat{\theta} - \theta_0)^T I(\theta_0)(\hat{\theta} - \theta_0)$$

$$S_1 = \{U(\hat{\theta})\}^T\{I(\theta_0)\}^{-1}\{U(\hat{\theta})\}$$

$$S_2 = \{U(\theta_0)\}^T\{I(\hat{\theta}_0)\}^{-1}\{U(\theta_0)\}$$

and

$$S_3 = \{U(\hat{\theta})\}^T\{I(\hat{\theta})\}^{-1}\{U(\hat{\theta})\}$$

For more detail on alternative forms and properties of the score statistic, see Bera and Meckenzie (1986).

In general L, W, and S have asymptotic χ_r^2 distributions, where r is the rank of the matrix involved in the quadratic form defining the Wald or score statistic, and is for L the difference between the number of parameters estimated under the full or unrestricted model (the union of the null and alternative models) and under the null hypothesis. Note that the development presented here uses the information that the parameter space under the null hypothesis contains just one point.

3.3 The Likelihood Ratio, Wald, and Score Tests for Composite Null Hypothesis

In applications it is unlikely that if the distribution is unknown, then the parameters of the hypothesized distribution are known. If this is the case in the examples of the previous section, then the Poisson mean in Example 3.1, and the normal mean and standard deviation in Example 3.2 will have to enter the problem as "nuisance" (unspecified) parameters.

To deal with this added complication, the theory of §3.2 can be modified so that $X_1, \ldots, X_n$ is a random sample from a continuous distribution with probability density function $f(x; \gamma)$, where the parameter vector γ is partitioned via $\gamma = (\theta^T, \beta^T)^T$. Sampling from a discrete population is completely analogous. In the partitioning of γ, θ is a $k \times 1$ vector of real parameters, $\theta \in \Theta$, and β is a $q \times 1$ vector of real nuisance parameters, $\beta \in B$. We wish to test $H_0: \theta = \theta_0$ against $K: \theta \neq \theta_0$ without specifying β; again θ_0 should be an interior point of Θ to avoid continuity problems. The logarithm of the likelihood is

$$l(\gamma; x) = \sum_{i=1}^{n} \log f(x_i; \gamma)$$

and the natural extension of L from §3.2 is

$$\hat{L} = 2l(\hat{\gamma}; X) - 2l(\hat{\gamma}_0; X)$$

where $\hat{\gamma} = (\hat{\theta}^T, \hat{\beta}^T)^T$ is the maximum likelihood estimator of γ under the full model, restricted only in that $\theta \in \Theta$ and $\beta \in B$; also $\hat{\gamma}_0^T = (\theta_0^T, \hat{\beta}_0^T)$ is the maximum likelihood estimator of γ under the null hypothesis, in which θ is restricted to taking the value θ_0 and β is restricted only in that $\beta \in B$.

The efficient score is $U(\gamma) = (\partial l(\gamma; X)/\partial \gamma_i)$ and the information matrix is $I(\gamma) = -E_\gamma[(\partial^2 l(\gamma; X)/\partial \gamma_i \partial \gamma_j)]$. Of course E_γ denotes expectation with respect to the distribution with probability density function $f(x; \gamma)$; E_0 denotes expectation with respect to the distribution with probability density function $f(x; \gamma_0)$. Now U and I may be partitioned as is γ, so that

$$U = U(\gamma) = \begin{pmatrix} U_\theta(\gamma) \\ U_\beta(\gamma) \end{pmatrix} \quad \text{and} \quad I = I(\gamma) = \begin{pmatrix} I_{\theta\theta}(\gamma) & I_{\theta\beta}(\gamma) \\ I_{\beta\theta}(\gamma) & I_{\beta\beta}(\gamma) \end{pmatrix}$$

Define $\Sigma(\gamma)$ by

$$\Sigma(\gamma) = I_{\theta\theta}(\gamma) - I_{\theta\beta}(\gamma) I_{\beta\beta}^{-1}(\gamma) I_{\beta\theta}(\gamma)$$

It follows from the discussion in Cox and Hinkley (1974, §9.3) that $\{\Sigma(\gamma)\}^{-1}$ is the asymptotic covariance matrix of $\hat{\theta}$ and that $\{\Sigma(\gamma)\}$ is the asymptotic covariance matrix of $U_\theta(\gamma)$. Then from Pearson's lemma (see §2.2), the statistics defined in the following have asymptotic χ^2 distributions. The first is the generalized Wald statistic,

$$\hat{W} = (\hat{\theta} - \theta_0)^T \Sigma(\hat{\gamma})(\hat{\theta} - \theta_0)$$

which requires only the unrestricted maximum likelihood estimator $\hat{\gamma}$. The second is the generalization of the score statistic, and is

$$\hat{S} = \{U_\theta(\hat{\gamma}_0)\}^T \{\Sigma(\hat{\gamma}_0)\}^{-1} \{U_\theta(\hat{\gamma}_0)\}$$

This requires the restricted maximum likelihood estimator $\hat{\gamma}_0$ and that the inverse is defined. Of course $\hat{L}$ requires both $\hat{\gamma}$ and $\hat{\gamma}_0$.

Although it is a slight digression, we will now discuss a classical problem that illustrates the techniques we have been examining.

Example 3.3.3. Behrens–Fisher Problem

In the Behrens–Fisher problem, $Y_1, \ldots, Y_m$ is a random sample from a $N(\mu_Y, \sigma_Y^2)$ population, and $Z_1, \ldots, Z_n$ is an independent random sample from a $N(\mu_Z, \sigma_Z^2)$ population. It is desired to test $H: \mu_Y = \mu_Z$ against $K: \mu_Y \neq \mu_Z$, with the standard deviations σ_Y^2, σ_Z^2 being nuisance parameters.

To conform to our notation, put $(Y_1, \ldots, Y_m, Z_1, \ldots, Z_n) = X^T$, $\mu_Y - \mu_Z = 2\theta$, $\mu_Y + \mu_Z = 2\beta_1$, $\sigma_Y^2 = \beta_2$, and $\sigma_Z^2 = \beta_3$. The likelihood is

$$\prod_{i=1}^{m} \frac{1}{\sigma_Y \sqrt{2\pi}} \exp\{-(y_i - \mu_Y)^2/(2\sigma_Y^2)\} \prod_{j=1}^{n} \frac{1}{\sigma_Z \sqrt{2\pi}} \exp\{-(z_j - \mu_Z)^2/(2\sigma_Z^2)\}$$

which has logarithm, in terms of θ and β,

$$l = \text{constant} - (m/2)\log\beta_2 - (n/2)\log\beta_3$$
$$- (2\beta_2)^{-1} \sum_i (y_i - \beta_1 - \theta)^2 - (2\beta_3)^{-1} \sum_j (z_j - \beta_1 + \theta)^2$$

Routine calculations give first order derivatives

$$\frac{\partial l}{\partial \theta} = \frac{\Sigma(y_i - \beta_1 - \theta)}{\beta_2} - \frac{\Sigma(z_j - \beta_1 + \theta)}{\beta_3} \tag{3.3.1}$$

$$\frac{\partial l}{\partial \beta_1} = \frac{\Sigma(y_1 - \beta_1 - \theta)}{\beta_2} + \frac{\Sigma(z_j - \beta_1 + \theta)}{\beta_3} \tag{3.3.2}$$

$$\frac{\partial l}{\partial \beta_2} = -\frac{m}{2\beta_2} + \frac{\Sigma(y_i - \beta_1 - \theta)^2}{2\beta_2^2}$$

$$\frac{\partial l}{\partial \beta_3} = -\frac{n}{2\beta_3} + \frac{\Sigma(z_j - \beta_1 + \theta)^2}{2\beta_3^2}$$

Under the full model we equate all four derivatives to zero. Equating the first two to zero leads to

$$\hat{\theta} = (\bar{Y} - \bar{Z})/2 \quad \text{and} \quad \hat{\beta}_1 = (\bar{Y} + \bar{Z})/2$$

Working with the third and fourth derivatives gives

$$\hat{\beta}_2 = \sum (Y_i - \bar{Y})^2/m, \qquad \hat{\beta}_3 = \sum (Z_j - \bar{Z})^2/n$$

Under the null hypothesis the situation is more complicated. The first equation is omitted and θ replaced by zero in the other equations. Equating Equation 3.3.2 to zero gives

$$m(\bar{y} - \hat{\beta}_{10})/\hat{\beta}_{20} + n(\bar{z} - \hat{\beta}_{10})/\hat{\beta}_{30} = 0$$

which may be reorganized to give $\hat{\beta}_{10}$ as a weighted mean of $\bar{Y}$ and $\bar{Z}$:

$$\hat{\beta}_{10} = \{\bar{Y}(m/\hat{\beta}_{20}) + \bar{Z}(n/\hat{\beta}_{30})\}/\{(m/\hat{\beta}_{20}) + (n/\hat{\beta}_{30})\} \qquad (3.3.3)$$

Substituting this in

$$\hat{\beta}_{20} = \sum (Y_i - \hat{\beta}_{10})^2/m \quad \text{and} \quad \hat{\beta}_{30} = \sum (Z_j - \hat{\beta}_{10})^2/n$$

leads to a cubic equation. Write $S_Y^2 = \Sigma (Y_i - \bar{Y})^2/(m-1)$ and $S_Z^2 = \Sigma (Z_j - \bar{Z})^2/(n-1)$; the cubic is

$$\begin{aligned}(m+n)\hat{\beta}_{10}^3 - \{(m+2n)\bar{Y} + (n+2m)\bar{Z}\}\hat{\beta}_{10}^2 \\ + \{n(m-1)S_Y^2/m + m(n-1)S_Z^2/n + 2(m+n)\bar{Y}\bar{Z} + n\bar{Y}^2 + m\bar{Z}^2\}\hat{\beta}_{10} \\ - \{m\bar{Y}[(n-1)S_Z^2/n + \bar{Z}^2] + n\bar{Z}[(m-1)S_Y^2/m + \bar{Z}^2]\} = 0\end{aligned}$$

In practice, the cubic is solved numerically. If there are three real roots, as sometimes occurs, the likelihood must be evaluated to see which one maximizes it. Having found $(\hat{\beta}_0)_1 = \hat{\beta}_{10}$, we easily have

$$\hat{\beta}_{20} = \sum (Y_i - \hat{\beta}_{10})^2/m \quad \text{and} \quad \hat{\beta}_{30} = \sum (Z_j - \hat{\beta}_{10})^2/n$$

Clearly this is one situation in which the Wald statistic looks distinctly more appealing than both $\hat{L}$ and $\hat{S}$, both of which require the cubic to be solved (see Bozdogan and Ramirez, 1986, who discussed solving the cubic).

To confirm that the solutions to the likelihood equations do maximize them, and to find the information matrix that is required to calculate $\hat{W}$ and $\hat{S}$, we need the second derivatives of l:

$$\frac{\partial^2 l}{\partial \theta^2} = -\frac{m}{\beta_2} - \frac{n}{\beta_3}, \qquad \frac{\partial^2 l}{\partial \theta\, \partial \beta_1} = -\frac{m}{\beta_2} + \frac{n}{\beta_3}, \qquad \frac{\partial^2 l}{\partial \theta\, \partial \beta_2} = -\frac{m(\bar{y} - \beta_1 - \theta)}{\beta_2^2}$$

$$\frac{\partial^2 l}{\partial \theta\, \partial \beta_3} = \frac{n(\bar{z} - \beta_1 + \theta)}{\beta_3^2}, \qquad \frac{\partial^2 l}{\partial \beta_1^2} = -\frac{m}{\beta_2} - \frac{n}{\beta_3}, \qquad \frac{\partial^2 l}{\partial \beta_1\, \partial \beta_2} = -\frac{m(\bar{y} - \beta_1 - \theta)}{\beta_2^2}$$

$$\frac{\partial^2 l}{\partial \beta_1\, \partial \beta_3} = -\frac{n(\bar{z} - \beta_1 + \theta)}{\beta_3^2}, \qquad \frac{\partial^2 l}{\partial \beta_2^2} = \frac{m}{2\beta_2^2} - \frac{\Sigma (y_i - \beta_1 - \theta)^2}{\beta_2^3}$$

$$\frac{\partial^2 l}{\partial \beta_2\, \partial \beta_3} = 0 \quad \text{and} \quad \frac{\partial^2 l}{\partial \beta_3^2} = \frac{n}{2\beta_3^2} - \frac{\Sigma (z_j - \beta_1 + \theta)^2}{\beta_3^3}$$

The information matrix has elements minus the expected value of these derivatives:

$$I(\gamma) = \begin{pmatrix} \frac{m}{\beta_2} + \frac{n}{\beta_3} & \frac{m}{\beta_2} - \frac{n}{\beta_3} & 0 & 0 \\ \frac{m}{\beta_2} - \frac{n}{\beta_3} & \frac{m}{\beta_2} + \frac{n}{\beta_3} & 0 & 0 \\ 0 & 0 & \frac{m}{2\beta_2^2} & 0 \\ 0 & 0 & 0 & \frac{n}{2\beta_3^2} \end{pmatrix}$$

This uses results such as $E[\bar{Y}] = \beta_1 + \theta$ and $E[\Sigma\,(Z_j - \beta_1 + \theta)^2] = n\beta_3$, which are true under both the restricted and unrestricted models. It follows that

$$\begin{aligned} \Sigma(\gamma) &= \left\{\frac{m}{\beta_2} + \frac{n}{\beta_3}\right\} \\ &\quad - \left(\frac{m}{\beta_2} - \frac{n}{\beta_3}, 0, 0\right) \operatorname{diag}\left\{\left(\frac{m}{\beta_2} + \frac{n}{\beta_3}\right)^{-1}, \frac{2\beta_2}{m}, \frac{2\beta_3}{n}\right\}\left(\frac{m}{\beta_2} - \frac{n}{\beta_3}, 0, 0\right)^T \\ &= \left(\frac{m}{\beta_2} + \frac{n}{\beta_3}\right) - \left(\frac{m}{\beta_2} - \frac{n}{\beta_3}\right)^2\left(\frac{m}{\beta_2} + \frac{n}{\beta_3}\right)^{-1} \\ &= 4mn/(m\beta_3 + n\beta_2) \end{aligned}$$

Hence

$$\begin{aligned} \hat{W} &= (\bar{Y} - \bar{Z})^2/\{\hat{\beta}_2/m + \hat{\beta}_3/n\} \\ &= (\bar{Y} - \bar{Z})^2/\{(m-1)S_Y^2/m^2 + (n-1)S_Z^2/n^2\} \end{aligned}$$

When $\theta = 0$ the efficient score is given by $m(\bar{Y} - \hat{\beta}_{10})/\hat{\beta}_{20} + n(\bar{Z} - \hat{\beta}_{10})/\hat{\beta}_{30}$. Using Equation 3.3.3, this becomes $2(\bar{Y} - \bar{Z})/\{\hat{\beta}_{20}/m + \hat{\beta}_{30}/n\}$. Now using $\Sigma(\gamma)$ from the preceding equation we have

$$\hat{S} = (\bar{Y} - \bar{Z})^2/\{\hat{\beta}_{20}/m + \hat{\beta}_{30}/n\}$$

$\hat{L}$, $\hat{W}$, and $\hat{S}$ all have null χ_1^2 distributions.

In large samples, $\hat{W}$ is hardly distinguishable from V^2 where V is due to Welch (1937) and is given by

$$V = (\bar{Y} - \bar{Z})/\sqrt{\{S_Y^2/m + S_Z^2/n\}}$$

A size and power study by Best and Rayner (1987b) recommended use of V in preference to $\hat{L}$ and $\hat{S}$ when critical points of V are determined from a t-distribution with estimated degrees of freedom. If this is done, the exact small sample sizes are remarkably close to those given by the approximating distribution. Other recent studies, such that of Scariano and Davenport (1986), have reached similar conclusions.

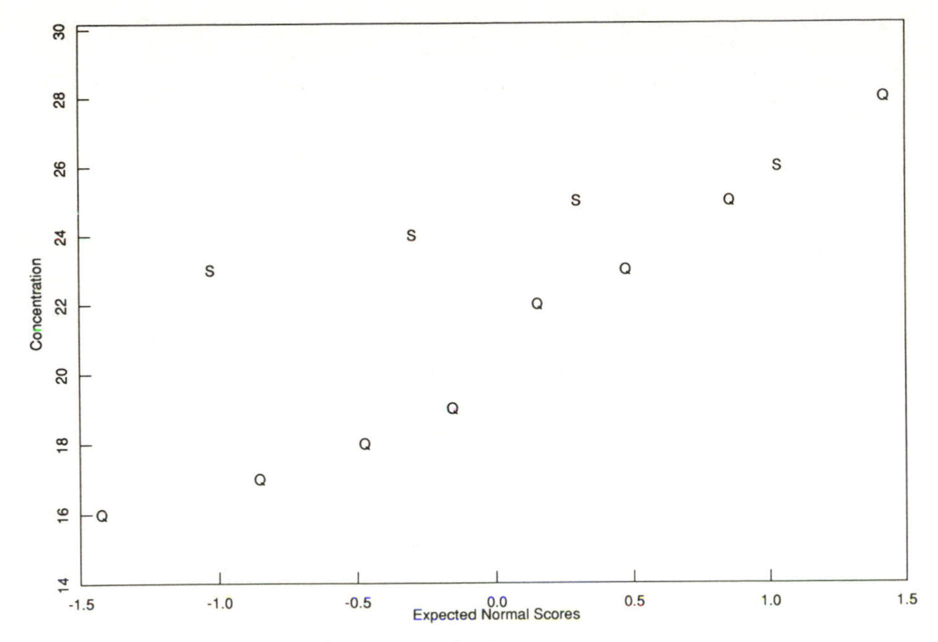

Figure 3.1 Normal Q–Q plot for chemical data.

Example 3.3.4

Consider the following two sets of data, which are determinations of the concentration of a chemical using a standard and a quick method:

S: 23, 24, 25, 26
Q: 23, 18, 22, 28, 17, 25, 19, 16

Normal Q–Q plots for each set of data can be superimposed and given in one figure (see Figure 3.1). The "by eye" lines through the S points and through the Q points would have different slopes. This indicates different standard deviations, so that the pooled t-test is inappropriate. The linearity of the plots for the S and Q data suggests that the normality assumption is plausible. Note that in §6.2 we develop a significance test to assess normality. The plot shows that overall the S values are higher than the Q values, indicating the S mean is higher than the Q mean. A significance test using V confirms this, although the pooled t-test does not. The P values are .096 for the pooled t-test and .036 for the test based on V.

3.4 Properties of the Asymptotically Optimal Tests

Subsequently, we will observe that Neyman's original smooth tests were constructed to satisfy a certain small sample optimality condition. Not only is this notion difficult to work with, it is difficult to extend to more useful (especially composite) situations. We therefore turn to the asymptotically optimal tests outlined in the two previous sections. These tests are generally available, and are certainly applicable to the models we propose. The mechanics of their application present no real difficulties; however, greater recommendations than these are needed to properly justify their use, and it is to this end that we now turn our attention.

There is some literature on the boundary problems. For example:

> The $C(\alpha)$ tests and asymptotic tests based on maximum likelihood estimators are equivalent when the null hypothesis implies that the set of all parameters, both those being tested and any incidental parameters, lie interior to an open set in the parameter space . . . even when only one of the parameters lies on its boundary, if the null hypothesis involves more than one parameter, the $C(\alpha)$ and maximum likelihood tests are no longer equivalent. (Moran, 1973, p. 80)

These difficulties do not arise in the problems we will discuss, and therefore were avoided in the initial discussion by the insistence that θ_0 be interior to Θ.

Javitz (1975) showed that the score statistic is asymptotically equivalent to the likelihood ratio statistic. In the absence of nuisance parameters, and for the single parameter case, Cox and Hinkley (1974) argued that the equivalent tests

> are asymptotically most powerful unbiased (p. 319)

having prefixed this by saying that

> only local alternatives are important for very large n (p. 318).

They also noted that no such test existed in the multiparameter case. This discussion is clarified by Gart and Tarone (1983, p. 781):

> for the case in which a single parameter is tested in the presence of nuisance parameters, the score test is simply the normal approximation to the uniformly most powerful unbiased (UMPU) test. Obviously, for one-parameter exponential families the score test is not only asymptotically locally most powerful but is also an approximation to the UMPU test

With regard to optimality, Bhat and Nagnur (1965) observed that the $C(\alpha)$ tests of Neyman were defined by him to be "locally asymptotically most powerful". Since this does not generalize to more than one dimension, they introduced the notion of a "locally asymptotically most stringent test". This is the optimality conferred by the multivariate $C(\alpha)$ test.

Since the likelihood ratio, Wald, and score tests are asymptotically equivalent, we must resolve the question of how to choose between them. Maximum likelihood estimators under either the null or unrestricted hypotheses may be difficult to calculate, and so eliminate all but one option. In general, however

> C. R. Rao conjectured that the test based on the score function is likely to be locally more powerful than the likelihood ratio and Wald's tests. It is shown

> here that a suitable modification of Rao's conjecture is true in one parameter family of distributions satisfying certain regularity conditions and that, of the three tests, Wald's is the worst. . . . It appears that these three tests perform equally well for families with small statistical curvature when local alternatives are considered. . . . these tests have the same Pitman efficiency. (Chandra and Joshi, 1983, p. 226)

> The unbiasedness up to $o(n^{-1/2})$ of the LR test . . . implies the same for the other tests . . . (Chandra and Joshi, 1983, p. 235)

So Pitman efficiency and unbiasedness do not provide a basis for distinguishing between the tests.

The confirmation of Rao's conjecture by Chandra and Joshi, albeit restricted, suggests careful use of Wald's test. This is supported on other grounds. Vaeth (1985) noted that, contrary to the likelihood ratio test, Wald's test is not invariant under reparameterization. It is shown in that article that with the use of the Wald test, according to the parametrization employed, that the same data can be both consistent with all possible null values for the parameter and inconsistent with all possible values. Mantel (1987) added

> The anomalous behaviors demonstrated by Vaeth (1985) . . . correspond to situations so incompatible with the null value that they make the Wald statistic completely unsuitable. . . . Use of parameter estimates in $\hat{V}(\hat{\beta}_j)$ may lead to reasonable estimates of $V(\hat{\beta}_j)$, but if the null hypothesis is grossly violated $\hat{V}(\hat{\beta}_j)$ may fail to approximate $V(\hat{\beta}_j)$ (pp. 147–48).

Bera and McKenzie (1986) discussed several different forms of the score test. These involve replacing the covariance matrix of the score vector by asymptotically equivalent matrices. Some of these equivalent forms were given in §3.2, and we note that there are corresponding equivalent forms for the composite case. Bera and McKenzie (1986) commented on the *invariance* of the score test. For them, this refers to the facts that the score test

1. Does not use maximum likelihood estimators under the alternative.
2. Has the same form for fairly general models.

In examples they have looked at, the invariance does not imply poor power.

Ronchetti (1987, p. 4, 5) studied the robustness of the $C(\alpha)$ test.

> Therefore, a test statistic with a bounded influence function guarantees robustness of validity and robustness of efficiency for the test. . . . Neyman's optimal $C(\alpha)$ test . . . , though asymptotically efficient, has an unbounded . . . influence function.

Note that the influence function being unbounded doesn't necessarily imply lack of the claimed robustnesses. Nevertheless, this may be a shortcoming of the $C(\alpha)$ and equivalent tests.

There is an expanding literature on the properties of the asymptotically optimal tests, not the least of which refers to small sample correction factors that improve the approximations to the null distributions. The approach we take subsequently, of empirically finding such factors, may ultimately be demonstrated to be inferior. That is just one of several matters for further investigation.

In general problems of hypothesis testing, the likelihood ratio test has for many years been the first attempt solution, the standard that new methods must improve upon. We suggest that now the score and Wald tests must stand alongside the likelihood ratio test, to be assessed at the same time. Our survey of the literature does not lead us to prefer any particular one. In our recent investigation of the Behrens–Fisher problem (Best and Rayner, 1987b), the Wald test seemed the most appropriate. In the smooth models of this monograph, the score test appears to be the most convenient.

4

Neyman Smooth Tests for Simple Null Hypotheses

4.1 Neyman's Ψ^2 Test

Just as Pearson's (1900) paper, which we discussed in §2.1, is the foundation of that branch of Statistics known as goodness of fit, so is Neyman's (1937) paper the foundation of the subbranch, the smooth tests of goodness of fit. In §1.2, the brief history of smooth goodness of fit tests was outlined. Although Neyman's test was not the first smooth test, the history of smooth goodness of fit tests began with Neyman (1937) because it was some time before it was appreciated that Pearson's X^2 test was a smooth test. We will show this in §5.1, but for now we will review Neyman (1937).

Neyman (1937) was dedicated to the memory of Karl Pearson and appeared shortly after Pearson's death in April 1936. It was also part of the series of papers in which Neyman and E. S. Pearson outlined their general philosophy of statistical inference, a philosophy that has had a profound effect on statistical inference.

The paper was presented much like a novelette. It started with a contents, which told us that the paper had four sections. The "Introductory" section contained the notation and terminology, and definitions of concepts such as *simple* and *composite* hypotheses, *critical region, level of significance, errors of the first and second kind,* and less formally, "*smooth*" *tests for goodness of fit.*" Of course, the basic definitions reflected the fact that statistical inference was very young at this time. Smooth alternative probability density functions were, in this introductory section, presented visually, in terms of probability density functions that have few intersections with the null probability density function. Later Barton (1953) would find fault with this interpretation of the formal definition presented in Neyman's §2.

The second section was entitled " 'Smooth' test for goodness of fit." It began by pointing out that by using the probability integral transformation, any completely specified continuous probability density function can be transformed to uniformity. It was sufficient, therefore, to consider tests for uniformity. Now "smooth" alternatives to uniformity were defined by

$$C(\Theta) \exp\left\{\sum_{i=1}^{k} \Theta_i \pi_i(y)\right\}, \qquad 0<y<1 \tag{4.1.1}$$

where $\{\pi_i(y)\}$ was the set of orthonormal polynomials on the uniform distribution (the normalized Legendre polynomials, defined in §1.2) and $C(\Theta)$ was a constant depending on Θ, introduced to ensure that the probability density function integrated to one. Of course, testing for uniformity was equivalent to testing H_0: $\Theta = 0$ against K: $\Theta \neq 0$. We might speculate on the form chosen in Equation 4.1.1. Fourier series or Gram–Charlier expansions may have suggested $\{\sum \Theta_i \pi_i(y)\}$; using it as an exponent avoids negative frequencies.

In subsection (c) Neyman then announced the optimality criterion to be applied to this problem. The test would be required to be, in modern terminology, locally uniformly most powerful symmetric, unbiased, and of size α for testing uniformity against the alternatives (see Equation 4.1.1). To achieve a solution, it was first necessary to transform the Θ_i by putting $\theta_i = \Theta_i \sqrt{n}$, $i = 1, \ldots, k$. Now what has become known as *The Generalized Fundamental Lemma of Neyman and Pearson* was invoked to show that the optimal test is *approximately* that which rejects for large values of

$$\Psi_k^2 = \sum_{i=1}^{k} U_i^2, \quad \text{in which} \quad U_i = \sum_{j=1}^{n} \pi_i(Y_j)/\sqrt{n}$$

What was actually shown was that the test that rejected for large values of Ψ_k^2 was most powerful subject to certain constraints. These constraints are not precisely unbiasedness, symmetry and so on. However asymptotically the joint distribution of the U_i is multivariate normal, and asymptotically the constraints are satisfied. As a consequence of these results, the asymptotic null distribution of Ψ_k^2 was χ_k^2.

Having derived the test, Neyman immediately applied it to some data of Mahalanobis (1934). The data were tested for consistency with the standard normal distribution, and with a beta distribution, using the test statistic Ψ_4^2. For the normal null, values of the test statistics were:

k	1	2	3	4
U_i^2	0.047	0.391	3.624	0.129
Ψ_k^2	0.047	0.438	4.062	4.191

In assessing these values, we claim that since U_r^2 is based on a polynomial of degree r, it reflects consistency of the data with the rth moment of the null distribution. Although none of the Ψ_k^2 are significant, U_3^2 is sufficiently large that a test to assess if the data were positively skewed is significant at the 5% level.

With the beta null, values of the test statistics are:

k	1	2	3	4
U_i^2	0.720	240.957	1.494	333.426
Ψ_k^2	0.720	241.677	243.171	576.597

Ψ_1^2 was not significant at the 5% level, but Ψ_2^2, Ψ_3^2, and Ψ_4^2 were. Looking at the components, it is clear that the variance and kurtosis are excessively large for the data to be consistent with the null (beta) hypothesis. The components U_i^2 are clearly more informative than the test statistics Ψ_i^2. Neyman was very clear about how the statistics should be used.

> It must be understood that for any given hypothesis tested and any given conditions specifying the set of alternatives, there exists just one single test which should be selected, as it were, in advance and applied at a chosen level of significance. Some authors calculate several criteria at the same level of significance to test a single hypothesis . . . . They sometimes (i) reject the hypothesis tested whenever at least one of these criteria appears to be unfavourable or alternatively, (ii) combine the calculated criteria to form some new resulting criterion . . . by the procedure (i) the original level of significance is altered . . . . Against both (i) and (ii) . . . these procedures invariably introduce a *new* critical region, usually of unknown properties which may be biassed or may have some other defects.

This is different from our more casual (data analytic) approach, in which, if the null hypothesis is rejected, the components are used informally to suggest the nature of the departure from the null hypothesis.

The third section contains a proof that the asymptotic power function of the Ψ_k^2 test is noncentral χ^2 with k degrees of freedom and parameter of noncentrality $n \sum_{i=1}^{k} \Theta_i^2 = \sum_{i=1}^{k} \theta_i^2$.

"Interpretation of the results obtained and general remarks" was the title of the fourth and final section. The first two subsections discussed using the asymptotic alternative distribution to determine either powers or sample sizes required to achieve specified powers. Then the effect of using Ψ_k^2 to detect $C(\Theta) \exp\{\Theta_1 \pi_1(y) + \ldots + \Theta_m \pi_m(y)\}$, $k \neq m$, was addressed. Although not phrased in these terms, the conclusion was that U_i^2 was a "detector" for Θ_i. A test statistic that excluded a particular U_r^2 cannot detect the corresponding Θ_r. Correspondingly, if the test statistic included a particular U_r^2 when the corresponding θ_r was not present in the alternative, then the resulting test would not be as powerful as the test that excluded the unnecessary detector.

Now consider general alternatives of the form of Equation 4.1.1. We may think of general alternatives being approximated by Equation 4.1.1-type alternatives, but more accurately such a general alternative can be projected into the space spanned by $\pi_1(y), \ldots, \pi_k(y)$, and *these projections* can be detected by Ψ_k^2. How successful this test will be depends on how representative these projections are of the full alternative. To detect general alternatives, Neyman felt that taking $k = 4$ or 5 in Ψ_k^2 would detect a sufficiently broad class.

4.2 Neyman Smooth Tests for Uncategorized Simple Null Hypotheses

Neyman derived the Ψ_k^2 criterion as the approximate solution to a small sample optimality problem. Some of the salient features of §4.1 are that the optimality is appropriate for all sample sizes, including small ones; and that the constraints are only satisfied in the limit, as the sample size tends to infinity. Unfortunately, it could also be claimed that the optimality, locally uniformly most powerful unbiased symmetric size α, is somewhat convoluted and unnatural.

Instead of adopting Neyman's approach, in this section the Ψ_k^2 criterion is derived as the score statistic for a model equivalent to Equation 4.1.1. Thus, the desirable large sample properties of score tests, equivalent to those of the likelihood ratio tests, apply to Ψ_k^2. These properties complement the small sample properties imposed by the optimality criteria used in Neyman's approach.

We hypothesize that we have a random sample $X_1, \ldots, X_n$ from a continuous population with probability density function $f(x)$ and cumulative distribution function $F(x)$. Both of these are completely specified, so there are no unspecified (nuisance) parameters. We could apply, as did Neyman, the probability integral transformation; however, the results are a little neater without applying it. They are also consistent with the later results for models that are discrete, and that involve nuisance parameters. The development for discrete distributions is entirely parallel. In terms of the description given in §1.3, we are dealing with completely specified uncategorized models.

An *order k alternative* probability density function is defined by

$$g_k(x) = C(\theta) \exp\left\{\sum_{i=1}^{k} \theta_i h_i(x)\right\} f(x)$$

where $\{h_i(x)\}$ is a set of orthonormal functions on $f(x)$, and so satisfies

$$\int_{-\infty}^{\infty} h_r(x) h_s(x) f(x)\, dx = \delta_{rs}, \qquad r \text{ and } s = 0, 1, 2, \ldots$$

The θ_j are real parameters; $\theta = (\theta_1, \ldots, \theta_k)^T$ and $C(\theta)$ is a normalizing constant inserted so the probability density function integrates to one. For convenience $h_0(x) = 1$ for all x, so that

$$\int_{-\infty}^{\infty} h_r(x) f(x)\, dx = \delta_{r0}, \qquad r = 1, 2, 3, \ldots$$

Note that for the remainder of this section, expectations and covariances with respect to the probability density function $g_k(x)$ are denoted by $E_k[\ ]$ and $\text{cov}_k(\ ,\ )$, respectively, while expectations and covariances with respect to the probability density function $f(x)$ are denoted by $E_0[\ ]$ and $\text{cov}_0(\ ,\ )$, respectively. Choosing $h_0(x)$ to be always 1 ensures that $E_0[h_r(X)] = \delta_{r0} = 0$ for all $r \geq 1$.

Testing for the hypothesized distribution is equivalent to testing $H_0: \theta = 0$ against $K: \theta \neq 0$. The score statistic for this model is now derived. The likelihood is

$$\prod_{j=1}^{n} \left\{C(\theta) \exp\left\{\sum_{i=1}^{k} \theta_i h_i(x_j)\right\} f(x_j)\right\} = \{C(\theta)\}^n \exp\left\{\sum_i \sum_j \theta_i h_i(x_j)\right\} \prod_j f(x_j)$$

so that the logarithm of the likelihood is

$$l = n \log C(\theta) + \sum_i \sum_j \theta_i h_i(x_j) + \sum_j \log f(x_j)$$

Now in order to differentiate l we need to be able to differentiate $C(\theta)$. This is achieved in the following lemma.

Lemma 4.2.1 The derivatives of $\log C(\theta)$ satisfy

$$-\frac{\partial \log C(\theta)}{\partial \theta_i} = E_k[h_i(X)]$$

and

$$-\frac{\partial^2 \log C(\theta)}{\partial \theta_i\, \partial \theta_j} = \text{cov}_k(h_i(X), h_j(X))$$

Proof Consider that as a valid probability density function, $g_k(x)$ satisfies

$$1 = \int_{-\infty}^{\infty} g_k(x)\, dx$$

Differentiate this with respect to θ_i and θ_j, assuming that the order of integration and differentiation may be interchanged. This is valid by a result in Lehmann (1986, p. 59). First, we obtain

$$0 = \int_{-\infty}^{\infty} \left[\frac{\partial C(\theta)}{\partial \theta_i} \exp\left\{\sum \theta_i h_i(x)\right\} f(x) + h_i(x) C(\theta) \exp\left\{\sum \theta_i h_i(x)\right\} f(x)\right] dx$$

$$= \int_{-\infty}^{\infty} \left[\frac{\partial \log C(\theta)}{\partial \theta_i} + h_i(x)\right] g_k(x)\, dx$$

$$= \frac{\partial \log C(\theta)}{\partial \theta_i} + E_k[h_i(X)],$$

as required. Differentiating again gives

$$0 = \frac{\partial^2 \log C(\theta)}{\partial\theta_i\,\partial\theta_j} + \int_{-\infty}^{\infty} h_i(x)\frac{\partial g_k(x)}{\partial\theta_j}\,dx$$

$$= \frac{\partial^2 \log C(\theta)}{\partial\theta_i\,\partial\theta_j} + \int_{-\infty}^{\infty} h_i(x)\left[\frac{\partial \log C(\theta)}{\partial\theta_j} + h_j(x)\right]g_k(x)\,dx$$

$$= \frac{\partial^2 \log C(\theta)}{\partial\theta_i\,\partial\theta_j} + E_k[h_i(X)\{h_j(X) - E_k[h_j(X)]\}]$$

again yielding the required result.

We are now in a position to differentiate the logarithm of the likelihood l and proceed to the score statistic. The result is stated as a theorem.

Theorem 4.2.1. The score statistic for testing $H_0: \theta = 0$ against $K: \theta \neq 0$ given a random sample from a distribution with probability density function $g_k(x)$ is

$$S_k = \sum_{i=1}^{k} U_i^2, \quad \text{in which} \quad U_i = \sum_{j=1}^{n} h_i(X_j)/\sqrt{n}$$

Proof Differentiating l with respect to θ_i gives

$$\frac{\partial l}{\partial\theta_i} = n\frac{\partial \log C(\theta)}{\partial\theta_i} + \sum_j h_i(x_j)$$

and then differentiating with respect to θ_j gives

$$\frac{\partial^2 l}{\partial\theta_i\,\partial\theta_j} = n\frac{\partial^2 \log C(\theta)}{\partial\theta_i\,\partial\theta_j}$$

Now using Lemma 4.2.1 and that $E_0[h_r(X)] = 0$, for $r = 1, 2, \ldots, \partial l/\partial\theta_i = \sum_j h_i(x_j)$. Write $U_i = \sum_j h_i(X_j)/\sqrt{n}$. The efficient score is then

$$U = \left(\sum_j h_i(X_j)\right).$$

The information matrix, evaluated at $\theta = 0$, is

$$-E_0\left[\left(\frac{\partial^2 l}{\partial\theta_r\,\partial\theta_s}\right)\right] = n(\text{cov}_0(h_r(X), h_s(X))) = nI_k$$

where I_k is the $k \times k$ unit matrix. This gives the required score statistic.

Note that the U_i in the theorem is not quite the ith element of the efficient score: it differs by a factor $\sqrt{n}$. Also S_k is *very* similar to Ψ_k^2; the only difference is that the normalized Legendre polynomials have been replaced by polynomials orthonormal with respect to $f(x)$. Clearly Ψ_k^2 is a particular case of S_k.

By applying the derivation used in the theorem to the same problem, but with $g_k(x)$ replaced by $C(\theta)\exp\{\theta_r h_r(x)\}f(x)$, we find that U_r^2 is the score statistic for testing $H_{r0}: \theta_r = 0$ against $K_r: \theta_r \neq 0$. So as Neyman showed, albeit in a different way, U_r^2 is a detector of θ_r. Also, the theorem showed that the information matrix is diagonal. Since this is the asymptotic covariance matrix of the efficient score vector, which is asymptotically multivariate normal (see Cox and Hinkley, 1974, Chapter 9), the U_r are asymptotically independent. So in as much as the asymptotic properties hold, each U_r^2 is a detector for the corresponding θ_r *and no other*.

Finally, note that the information matrix, nI_k, has rank k, so that the asymptotic null and alternative distributions of S_k are χ_k^2; if $g_k(x)$ holds, the parameter of noncentrality is $n(\theta_1^2 + \ldots + \theta_k^2)$.

The advantage of S_k over Ψ_k^2 is that the orthonormal system can be chosen to give good power against particular alternatives. Everything that has been said here about continuous distributions readily carries over to discrete distributions.

4.3 Effective Order and Power Comparisons

For the present, let us concentrate on Neyman's Ψ_k^2 test for uniformity. Note, however, that the discussion can be easily modified to cover tests for other distributions, and indeed, all the smooth tests discussed in this monograph.

The test based on Ψ_k^2 was constructed to be "optimal" against alternatives with the density of Equation 4.1.1. "Optimal" may be interpreted either in Neyman's sense as discussed in §4.1, or in the sense that score statistics are optimal, as discussed in §3.4. The θ_i used to define the order k alternative to uniformity can be thought of as the coordinates of the alternative in a k dimensional alternative space spanned by the normalized Legendre polynomials $\{\pi_i(y)\}$.

Now the test is intended to be an omnibus test, suitable for detecting arbitrary alternatives, not just those of order k. If Ψ_k^2 is used when the alternative is not of order k, the projection of the alternative into the k dimensional space spanned by the k Legendre polynomials $\pi_1(y), \ldots, \pi_k(y)$ is detected.

The effect of using Ψ_{k+1}^2 rather than Ψ_k^2 depends on the alternative. The tests using Ψ_k^2 are equally sensitive to the first k components of the alternative, but completely insensitive to all other components. Of course Ψ_{k+1}^2 will be sensitive to the $k+1$st component, but since it is equally sensitive to all $k+1$ components, it will be less sensitive to the first k components than Ψ_k^2 was. In passing from Ψ_k^2 to Ψ_{k+1}^2 there is both a loss and a gain, a loss in sensitivity to each of the first k components, but a gain in sensitivity to the $k+1$st. The loss in sensitivity is sometimes known as *dilution*. (For example, see Kendall and Stuart, 1973, Chapter 30).

Whether or not Ψ_k^2 is more powerful than Ψ_{k+1}^2 for a particular alternative depends very much on the relative magnitudes of the θ_i. For known alternatives it is possible to use Fourier methods to calculate the θ_i, but for data it is not. So we have adopted the practice of applying a sequence of Ψ_k^2 tests to particular alternatives. Typically the power increases to a unique maximum and then decreases. If the most powerful level α Ψ_k^2 test occurs when $k = k^*$, we say that alternative has *level* α Ψ_k^2 *effective order* k^*. Our experience has been that the k^* is reasonably independent of the level α, of whether the test statistic is Ψ_k^2 or Pearson's X_P^2, and whether we use power or asymptotic relative efficiency (ARE) to find the most desirable test.

Of course the same procedure can be applied to data. A sequence of Ψ_k^2 tests can be applied to a data set to identify the effective order of the alternative generating the data. This would identify the best test to use on subsequent data.

Our experience in Best and Rayner (1985a) and Rayner, Best, and Dodds (1985) is that many of the alternatives used in power studies in the statistical literature have low effective order. To detect such alternatives, Ψ_k^2 with $k \leq 4$ will usually suffice. It is interesting that Miller and Quesenberry (1979) found Ψ_k^2 to be most effective in detecting a range of obviously low effective order alternatives. Unfortunately, they inappropriately applied X_P^2 tests of relatively high order, and so found X_P^2 to not be as effective as could be when appropriately applied. In §5.2, a report is given of Rayner, Best, and Dodds (1985), in which a detailed investigation of effective order was carried out.

4.4 Examples

Example 4.4.1. Birth-time Data

These data were previously analyzed in Example 1.4.2. There, it was commented that a test that did not group the data may be more effective at detecting an apparent trend in the data, but that the Kolmogorov–Smirnov test applied by Mood et al. (1979, p. 510) failed to do so.

The Neyman test of §4.1 gives $\Psi_4^2 = 5.36$ with components $U_1 = -0.16$, $U_2 = -1.54$, $U_3 = 0.48$, and $U_4 = 1.65$. The overall Ψ_4^2 value is again not significant, but we note that $(U_2^2 + U_4^2)/\Psi_4^2 > 0.95$. In a pilot study we would suggest this indicates there may be evidence of a departure from uniformity and further, that the departure is due to differences in variance and kurtosis rather than in mean or skewness. An appropriate test statistic for assessing subsequent data would be $U_2^2 + U_4^2$.

In the present context, a 5% level test for kurtosis greater than the uniform is just significant. It is invalid, however, to apply this test *after having viewed the data.* A possible explanation of the extra daytime births is that the data include all births, including those artificially induced. A doctor is possibly more likely to choose daytime for this procedure. If this had been anticipated before sighting the data, and it had been decided to apply the one-sided test mentioned earlier, then it may be judged that the kurtosis is too great to be consistent with uniformity.

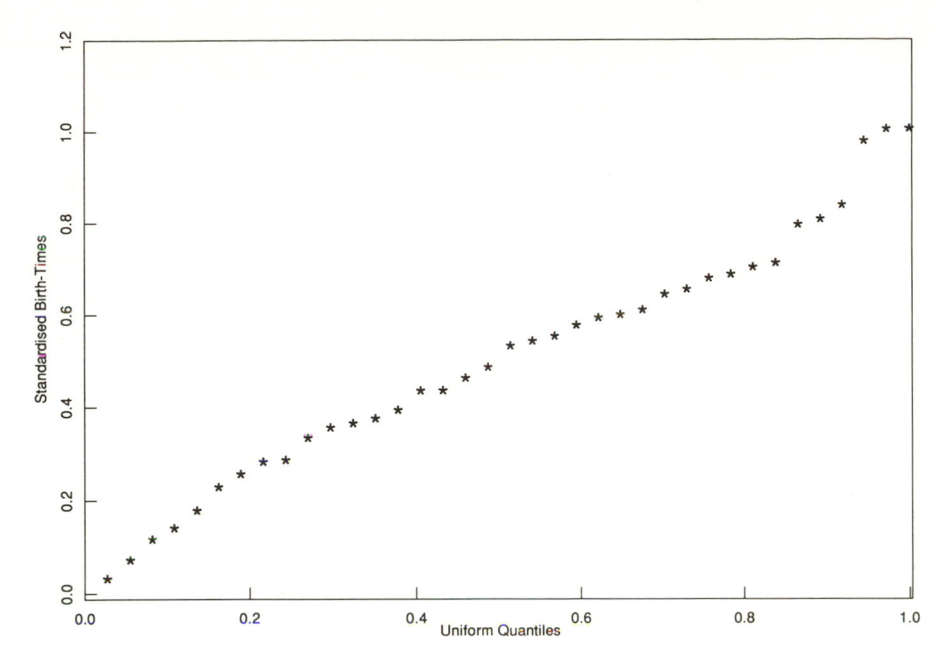

Figure 4.1 Uniform Q–Q plot of birth-time data.

Our own preference is to conclude that although this data do not give decisive evidence against uniformity, a larger data set may well do so. If such a set is gathered, it should be assessed for variance and kurtosis departures by calculating $U_2^2 + U_4^2$.

At the conclusion of §1.1 we remarked that graphic methods should be used alongside the significance tests we are developing here. So far we have done this with little comment. Although there are many methods from which to choose, just two are given here. The Q–Q plot of Figure 4.1 leaves us in some doubt. Is the plot sufficiently linear to confirm uniformity? Or is it sufficiently nonlinear to reject that hypothesis? Does the plot give any other information about the data?

Two graphic aids to the interpretation of Q–Q plots which we will not consider here are the Monte Carlo envelope of Atkinson (1985, p. 35) and the confidence bands of Michael (1983). Instead of illustrating these we show in Figure 4.2 a Legendre series estimate which is closely related to the components of the Neyman smooth test.

The Legendre series density estimate is simple to construct and interpret. Having chosen $k = 4$ in Equation 1.2.2, we simply replace the θ_i by their maximum likelihood estimates, the corresponding U_i. A better fit is obtained by not ignoring U_i for θ_i consistent with zero. Like the histogram for this data (Figure 1.2), the Legendre series density estimate shows the increased frequency in the middle of the day. An alternative approach is to use Equation 4.1.1, replacing the θ_i by their maximum likelihood estimates.

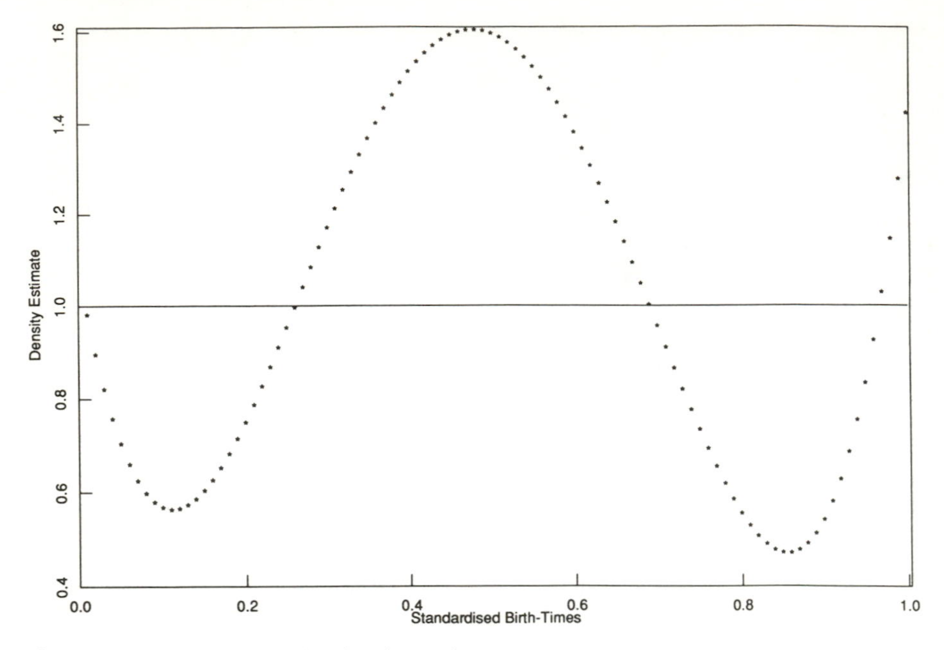

Figure 4.2 Legendre series density estimate.

Example 4.4.2. Combination of *P* Values

When a number of independent tests of significance have been made, and *P* values of $p_1, \ldots, p_n$ say obtained, Fisher (1925, p. 99) claimed that

> it sometimes happens that although few or none can be claimed individually as significant, yet the aggregate gives an impression that the probabilities are on the whole lower than would often have been obtained by chance.

Thus, suppose that as in Fisher (1925, Example 14.1), $p_1 = .145$, $p_2 = .263$ and $p_3 = .087$. Fisher's classical test takes $-2\{\log p_1 + \ldots + \log p_n\}$ and compares this with critical values of the χ^2_{2n} distribution. For the present example a combined *P* value from this test is about .075. Pearson (1938) suggested using $-2\{\log(1-p_1) + \ldots + \log(1-p_n)\}$, and this gives a combined *P* value of less than .025.

As we are essentially looking at a mean shift, it makes sense, for example, to look at the first component of Neyman's smooth test statistic, or equivalently $P^* = p_1 + \ldots + p_n$. The distribution of P^* is known exactly (see Stephens, 1966, p. 235). He obtained

$$P(P^* \leq z) = \sum_{j=0}^{m} (-1)^j \, {}^nC_j (z-j)^n \Big/ n!$$

where m is the integer such that $m \leq z < m + 1$. For Fisher's example $n = 3$,

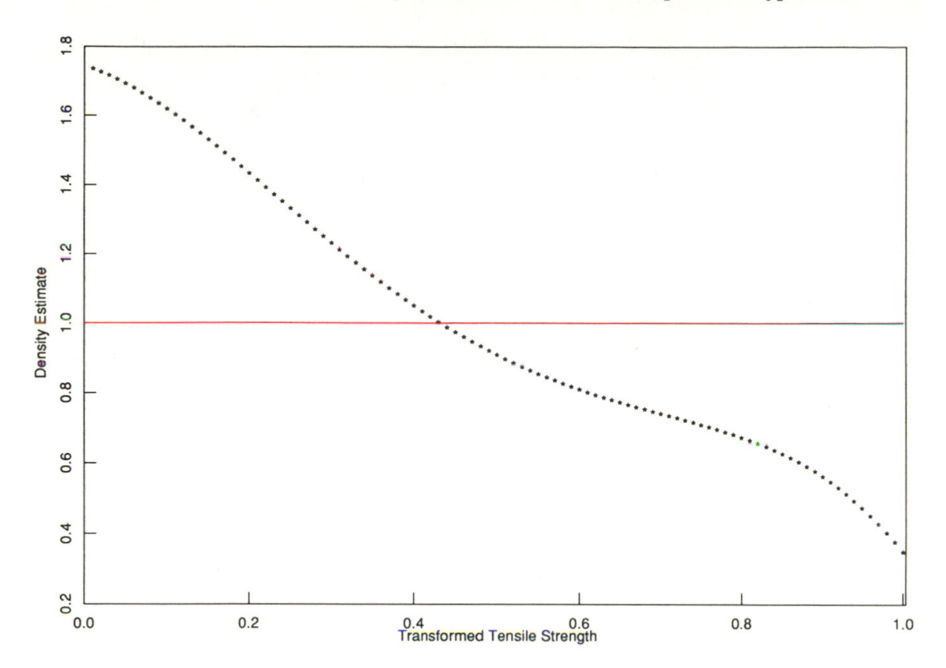

Figure 4.3 Legendre series density estimate.

$z = 0.495$, and $m = 0$, so that

$$P(P^* \leq z) = z^3/6 = 0.02$$

Example 4.4.3. Tensile Strength of Polyester Fibers

Quesenberry and Hales (1980) analyzed 32 measurements of the tensile strength of polyester fibers to see if they were consistent with the lognormal distribution. After a preliminary transformation, 30 observations were obtained. The data now follow.

0.023	0.032	0.054	0.069	0.081	0.094
0.105	0.127	0.148	0.169	0.188	0.216
0.255	0.277	0.311	0.361	0.376	0.395
0.432	0.463	0.481	0.519	0.529	0.567
0.642	0.674	0.752	0.823	0.887	0.926

If the lognormal hypothesis was correct, these observations should have been uniformly distributed. Although graphic evidence cast doubt on a uniformity hypothesis, formal significance tests could not reject it at the 10% level. In particular, Quesenberry and Hales obtained $\Psi_4^2 = 6.62$.

Looking more closely at the data, we find Ψ_4^2 has components $U_1 = -2.55$, $U_2 = 0.32$, $U_3 = -0.08$, and $U_4 = -0.17$. There is clearly a mean shift that Ψ_4^2 has been unable to detect because U_2, U_3, and U_4 have reduced or diluted its sensitivity. Figure 4.3 gives a Legendre series density estimate (see Example 4.4.1) for the transformed data where the mean shift shows up clearly as a trend. We remark here that examination of the components of Neyman's test for uniformity seems to have often been ignored.

This example highlights a difficulty with what we might call the *Quesenberry approach*—transforming the data so that a uniformity test is required to test the null hypothesis. The hypothesis has been rejected here, but what can we say about the original data? Our approach enables us to say something worthwhile about the transformed data; its corresponding probability density function may reasonably be modeled by Equation 1.2.2 with $k = 1$, since Θ_2, Θ_3 and Θ_4 may be taken to be zero. Our preference is to work as far as possible with the untransformed data so that if the null hypothesis is rejected, a ready alternative model is available.

5

Neyman Smooth Tests for Categorized Simple Null Hypotheses

5.1 Smooth Tests for Completely Specified Multinomials

The previous chapter dealt with uncategorized distributions with no nuisance parameters. Here the distribution is again completely specified, but the data are categorized into a finite number of categories. In §4.2 Ψ_k^2 was derived as the score test for testing $H_0: \theta = 0$ against $K: \theta \neq 0$ given a random sample from a population with probability density function $g_k(x)$. In exactly the same way, we may derive X_P^2 as a score test for the cell probabilities $p_1, \ldots, p_m$. This is done by imbedding the null probabilities in the smooth alternative cell probabilities

$$\pi_j = C(\theta) \exp\left\{\sum_{i=1}^{k} \theta_i h_{ij}\right\} p_j, \qquad j = 1, \ldots, m \qquad (5.1.1)$$

and testing $H_0: \theta = 0$ against $K: \theta \neq 0$. In terms of the classification of §1.3, we are dealing with a completely specified categorized model. In Equation 5.1.1, the θ_i, $i = 1, \ldots, k$ are real parameters, $\theta = (\theta_1, \ldots, \theta_k)^T$ and $C(\theta)$ is a normalizing constant so that $\sum_j \pi_j = 1$. Since the parameter space has dimension $m - 1$, we must have $k \leq m - 1$. For each i, $i = 1, \ldots, k$, the h_{ij} are values taken by a random variable H_i with $P(H_i = h_{ij}) = \pi_j$, $j = 1, \ldots, m$, for $i = 1, \ldots, k$. To obtain X_P^2 further restrictions on the h_{ij} must be imposed. We will now proceed with the derivation.

A random sample of n observations is taken and N_j, the number of observations in the jth cell, $j = 1, \ldots, m$, is noted. Write n_j for a realization of N_j, and note that $n = \sum_j n_j = \sum_j N_j$. Before finding the score statistic for the model given by Equation 5.1.1, some derivatives will be required. All expectations are assumed to exist.

Lemma 5.1.1. For each $r = 1, \ldots, k$, suppose H_r is a random variable that takes the value h_{rj} with probability π_j, $j = 1, \ldots, m$. Then

$$\frac{\partial \log \pi_j}{\partial \theta_r} = h_{rj} - E_k[H_r] \tag{5.1.2}$$

$$\frac{\partial^2 \log \pi_j}{\partial \theta_r \, \partial \theta_s} = \frac{\partial^2 \log C(\theta)}{\partial \theta_r \, \partial \theta_s} \tag{5.1.3}$$

$$\frac{\partial \log C(\theta)}{\partial \theta_r} = -E_k[H_r] \tag{5.1.4}$$

$$\frac{\partial^2 \log C(\theta)}{\partial \theta_r \, \partial \theta_s} = -\text{cov}_k(H_r, H_s) \tag{5.1.5}$$

Proof. The logarithm of π_j is

$$\log \pi_j = \log C(\theta) + \sum_j \theta_i h_{ij} + \log p_j$$

Differentiating gives

$$\frac{\partial \log \pi_j}{\partial \theta_r} = \frac{\partial \log C(\theta)}{\partial \theta_r} + h_{rj} = \frac{1}{\pi_j} \frac{\partial \pi_j}{\partial \theta_r} \tag{5.1.6}$$

and Equation 5.1.3. Now $\sum_j \pi_j = 1$, so that

$$\sum_j \partial \pi_j / \partial \theta_r = \sum \{\partial \log \pi_j / \partial \theta_r\} \pi_j = 0.$$

Using Equations 5.1.6 in this equation gives

$$\begin{aligned} 0 &= \sum \pi_j \left\{ \frac{\partial \log C(\theta)}{\partial \theta_r} + h_{rj} \right\} \\ &= \frac{\partial \log C(\theta)}{\partial \theta_r} + \sum h_{rj} \pi_j \\ &= \frac{\partial \log C(\theta)}{\partial \theta_r} + E_k[H_r] \end{aligned} \tag{5.1.7}$$

Equation 5.1.4 follows. Substituting this in Equation 5.1.6 gives Equation 5.1.2. Now differentiating Equation 5.1.4 with respect to θ_s gives

$$\begin{aligned} 0 &= \frac{\partial^2 \log C(\theta)}{\partial \theta_r \, \partial \theta_s} + \sum_j h_{rj} \frac{\partial \pi_j}{\partial \theta_s} \\ &= \frac{\partial^2 \log C(\theta)}{\partial \theta_r \, \partial \theta_s} + \sum_j h_{rj} \left[\frac{\partial \log C(\theta)}{\partial \theta_s} + h_{sj} \right] \pi_j \end{aligned}$$

using Equation 5.1.6. Now using Equation 5.1.4 and

$$\sum_j h_{rj}(h_{sj} - E_k[H_s]) \pi_j = \text{cov}_k(H_r, H_s)$$

Equation 5.1.7 follows.

We are now in a position to derive the score test. First, define the $m \times 1$ vectors $\pi = (\pi_r)$, $p = (p_r)$, and $N = (N_r)$, and the $k \times m$ matrix $H = (h_{ij})$.

Theorem 5.1.1. The score statistic for testing $H_0: \theta = 0$ against $K: \theta \neq 0$ for Equation 5.1.1 is

$$S_k = \{N - np\}^T H^T \Sigma^{-1} H\{N - np\}/n,$$

provided the $k \times k$ matrix $\boldsymbol{\Sigma}$, whose rsth element is $\text{cov}_0(H_r, H_s)$, is nonsingular.

Proof. The likelihood is the multinomial, and has logarithm

$$l = \text{constant} + \sum_{j=1}^{m} n_j \log \pi_j$$

The derivatives of l with respect to θ_r are

$$\begin{aligned} \frac{\partial l}{\partial \theta_r} &= \sum_{j=1}^{m} n_j \frac{\partial \log \pi_j}{\partial \theta_r} \\ &= \sum_{j=1}^{m} n_j \{h_{rj} - E_k[H_r]\} \quad \text{(using Equation 5.1.2)} \\ &= \sum_{j=1}^{m} n_j h_{rj} - nE_k[H_r] = (HN - nH\pi)_r \end{aligned}$$

because $(H\pi)_r = \sum_j h_{rj}\pi_j = E_k[H_r]$. Evaluated at $\theta = 0$, this yields the efficient score

$$U = H\{N - np\}$$

The second-order derivatives satisfy

$$\frac{\partial^2 l}{\partial \theta_r \, \partial \theta_s} = \sum_j n_j \frac{\partial^2 \log \pi_j}{\partial \theta_r \, \partial \theta_s} = n \frac{\partial^2 \log C(\theta)}{\partial \theta_r \, \partial \theta_s} = -n \, \text{cov}_k(H_r, H_s)$$

using Equations 5.1.3 and 5.1.5. It follows that when $\theta = 0$, the information matrix is $n\Sigma$. So the score statistic is as stated.

Theorem 5.1.2. If, in addition to the assumptions of Theorem 5.1.1, the h_{rs} satisfy for $r, s = 1, \ldots, m$

$$\sum_{j=1}^{m} h_{rj} h_{sj} p_j = \delta_{rs}, \quad \text{with} \quad h_{mj} = 1, \quad j = 1, \ldots, m, \tag{5.1.8}$$

then the score statistic when $k = m - 1$ is

$$S_{m-1} = X_P^2 = \sum_{j=1}^{m} (N_j - np_j)^2/(np_j)$$

Proof. Write $H^* = (h_{ij})$ for the $m \times m$ matrix H augmented by the mth row as defined. The orthonormality implies that

$$E_0[H_r] = (Hp)_r = \sum_{j=1}^{m} h_{rj}p_j = \delta_{rm} = 0$$

So we have $(\Sigma)_{rs} = \text{cov}_0(H_r, H_s) = \Sigma_j h_{rj}h_{sj}p_j - E_0[H_r]E_0[H_s] = \delta_{rs}$, and $\Sigma = I_{m-1}$, where I_k means the $k \times k$ identity matrix. Therefore, from Theorem 5.1.1, the score statistic is $S_{m-1} = \{N - np\}^T H^T H\{N - np\}/n$.

Now the orthonormality may be expressed as $H^* \,\text{diag}(p_i)H^{*T} = I_m$. This implies that $\text{diag}\,(p_i^{-1}) = H^{*T}H^* = H^TH + 11^T$, where 1 is the $m \times 1$ vector of ones. This gives

$$S_{m-1} = (N - E[N])^T\{\text{diag}(p_i^{-1}) - 11^T\}(N - E[N])/n$$
$$= \sum_{i=1}^{m} (N_i - np_i)^2/(np_i) - \left\{\sum_{i=1}^{m} (N_i - np_i)\right\}^2 \Big/ n = X_P^2$$

Corollary 5.1.1. X_P^2 may be partitioned into components, via $X_P^2 = V_1^2 + V_2^2 + \ldots + V_{m-1}^2$, where the V_r are asymptotically independent standard normal, and are given by $V_r = \Sigma_j h_{rj}N_j/\sqrt{n}$.

Proof. Define the vector of components by

$$V = (V_r) = H(N - np)/\sqrt{n} = HN/\sqrt{n}.$$

Then $S_{m-1} = V^TV = V_1^2 + \ldots + V_{m-1}^2$. Although the elements of N are correlated, V is asymptotically distributed as $m - 1$ variate normal, with asymptotic covariance matrix I_{m-1}, and so the V_r are asymptotically independent standard normal.

What components might be useful? First, note that if $L = (l_{rs})$ is any $m \times m$ orthogonal matrix with last row $(\sqrt{p_1}, \ldots, \sqrt{p_m})$ then $H^* = LD^{-1/2}$ satisfies Equation 5.1.8 with

$$V_r = \sum_{j=1}^{m} l_{rj}\{N_j/\sqrt{(np_j)}\}$$

The $\{N_j/\sqrt{(np_j)}\}$ represent the relative excess of the observed over the expected, and are natural quantities to consider in a goodness of fit context. Now the l_{rs} can be chosen to satisfy the orthogonality requirement and to give helpful comparisons. So if $l_{r1} = \sqrt{(p_1/(p_1 + p_2)\}} = -l_{r2}$ then V_r will compare the relative excess of observed over expected for cell 1 with that for cell 2. This will be further developed in later examples.

We will now put these results in their historical context. Barton (1955) modified Neyman's Ψ_k^2 test to deal with categorized data. If the multinomial distribution induced by categorizing has g classes and the alternative is of order k, then Barton's test is based on a statistic, say $B(g, k)$. If $g \to \infty$ in such a way that all class probabilities approach zero, and if the sample size $n \to \infty$, then $B(g, k) \to \Psi_k^2$. Also, $B(g, g - 1) = X_P^2$, the Pearson X^2 statistic based on g classes. Thus, if g is large, Pearson's X^2 test based on g classes can be regarded as a categorized smooth test of order $g - 1$. Kendall and Stuart (1973, §30.44) show that the $B(g, k)$ are in fact partitions of X_P^2. More recently, Kopecky and Pierce (1979, §3) stated that X_P^2 results from calculating a score statistic and using indicator functions of the g classes. None of these authors quite use the approach of this section, but the final message is perhaps the same.

5.2 X^2 Effective Order

The derivation of X_P^2 in the previous section shows that if the data are contained in m categories, then X_P^2 provides an order $m-1$ test. This agrees with Barton (1955). Since X_P^2 is an order $m-1$ test, the discussion in §4.3 on effective order for Ψ_k^2 tests can be applied equally to tests based on X_P^2.

We will now repeat part of a study by Rayner, Best and Dodds (1985) on the determination of effective order. The 11 alternatives listed in Table 5.1 were each successively taken as alternatives to uniformity, and approximate powers of X_P^2 calculated for m equally likely classes, $m = 2, 3, 4, \ldots, 50$,

Table 5.1 Some alternatives to uniformity

Alternative	Source*/Name	Probability density function ($f(y)$)	Nonzero domain
(1)	S/A, $k=1.5$	$1.5\surd(1-y)$	$0 \le y \le 1$
(2)	S/A, $k=2$	$2(1-y)$	$0 \le y \le 1$
	B&R/H_1, $m=2$		
(3)	S/B, $k=1.5$	$1.5\surd(2y)$	$0 \le y \le 1/2$
		$1.5\surd(2(1-y))$	$1/2 < y \le 1$
(4)	S/B, $k=2$	$4y$	$0 \le y \le 1/2$
	B&R/H_2, $m=2$	$4(1-y)$	$1/2 < y \le 1$
	Q&M/H_{21}		
(5)	S/B, $k=3$	$12y^2$	$0 \le y \le 1/2$
		$12(1-y)^2$	$1/2 < y \le 1$
(6)	S/C, $k=1.5$	$1.5\surd(1-2y)$	$0 \le y \le 1/2$
		$1.5\surd(3y-1)$	$1/2 < y \le 1$
(7)	S/C, $k=2$	$2(1-2y)$	$0 \le y \le 1/2$
	Q&M/H_{41}	$2(2y-1)$	$1/2 < y \le 1$
(8)	B&R/H_3, $m=2$	$2-6y$	$0 \le y \le 1/3$
		$6y-2$	$1/3 < y \le 2/3$
		$6-6y$	$2/3 < y \le 1$
(9)	B&R/H_4, $m=2$	$8y$	$0 \le y \le 1/4$
		$4-8y$	$1/4 < y \le 1/2$
		$8y-4$	$1/2 < y \le 3/4$
		$8-8y$	$3/4 < y \le 1$
(10)	Q&M/H_{22}	$1/\{2\surd y\}$	$0 \le y \le 1$
(11)	Q&M/H_{31}	$[\surd y + \surd(1-y)]/4$	$0 \le y \le 1$

* The alternatives to uniformity have been considered by S = Stephens (1974a), Q&M = Quesenberry and Miller (1977), and B&R = Best and Rayner (1981).

Table 5.2 Approximate powers* for the X_P^2 test with m equiprobable classes, with $n=20$, $m=2(1)10$, and $\alpha=0.5$

	m								
Alternative	2	3	4	5	6	7	8	9	10
(1)	**.258**	.236	.214	.196	.182	.170	.161	.153	.146
(2)	**.609**	.578	.537	.499	.466	.438	.413	.392	.374
(3)	.050	.164	.170	**.171**	.164	.158	.152	.146	.141
(4)	.050	**.456**	.441	.453	.428	.415	.394	.379	.362
(5)	.050	**.892**	.814	.868	.833	.839	.813	.807	.786
(6)	.050	**.205**	.170	.183	.164	.164	.152	.149	.141
(7)	.050	**.456**	.441	.453	.428	.415	.394	.379	.362
(8)	.116	.050	.361	**.384**	.363	.380	.370	.352	.347
(9)	.050	.143	.050	.302	.323	**.335**	.315	.332	.325
(10)	.457	.541	.582	.605	.619	.627	.632	.635	**.637**
(11)	.050	.115	.148	.168	.181	.191	.197	.202	**.205**

* The maximum power against each alternative is bold.

using the noncentral χ^2 approximation to the nonnull distribution of X_P^2. The results are given in Table 5.2.

As in §4.3, if for a particular alternative the most powerful level 5% test occurs for $m=m^*$ equiprobable classes, then we say that alternative has *level 5% X_P^2 effective order* m^*-1. Although such alternatives do not have the form of Equation 5.1.1 with $k=m^*-1$, they behave as though they do because we would expect the alternatives to Equation 5.1.1 to have greatest power for an order k test. The approximate effective orders are thus as given in Table 5.3.

The point is not that we now know what X_P^2 tests to apply to best detect these alternatives, for X_P^2 is an omnibus test, used when particular alternatives are not specified. If they were specified, more powerful directional tests could be applied. Rather, we know what alternatives particular X_P^2 tests best detect. So the equiprobable X_P^2 test with 10 classes is an inferior test to use to detect low-order alternatives such as alternatives (1) to (8). Our experience is that low-order alternatives occur more frequently than high-order alternatives, so that unless there is some reason to suspect high-order alternatives, a low-order test should be used.

For discrete alternatives of low order, if an X_P^2 test is to be used, it should be one with, say, 2, 3, or 4 classes. This will apparently waste information by overcategorization. If the extra calculation is feasible, we would

Table 5.3 Effective orders of Table 5.1 alternatives to uniformity

Effective Order	1	2	4	6	∞
Alternative	(1),(2)	(4),(5),(6),(7)	(3),(8)	(9)	(10),(11)

recommend using as many categories as possible, and use the first few *components* as described in the next section.

5.3 Components of X_P^2

In §5.1, the components V_r of X_P^2 were defined in terms of $\{h_{rs}\}$ that satisfy Equation 5.1.8. Many such $\{h_{rs}\}$ exist, and in a later example we will use elements of a Helmert matrix because these permit relevant comparisons for some data. However, we will now consider in detail components defined for the equiprobable situation by using the Chebyshev orthogonal polynomials given, for example, in Fisher and Yates (1963, Table 23) or in Pearson and Hartley (1970, Table 47). We have

$$X_P^2 = \sum_{r=1}^{m-1} V_r^2, \quad \text{where} \quad V_r = \sum_{i=0}^{m-1} f_r(i) X_{i+1} / \sqrt{(ns_r)}$$

in which $s_r = \sum_{i=0}^{m-1} f_r^2(i)/m$. The $f_r(i)$ may be read from the tables, as illustrated in Best and Rayner (1982), or calculated from the recurrence relation

$$f_{r+1}(i) = f_1(i) f_r(i) - \alpha_r f_{r-1}(i)$$

with $f_0 = 1$, $f_1(i) = i - (m-1)/2$ and $\alpha_r = r^2(m^2 - r^2)/\{4(4r^2 - 1)\}$.

The procedure we recommend in practice is to calculate $V_1, V_2, \ldots, V_r$ and the residual $X_P^2 - V_1^2 - \ldots - V_r^2$. The asymptotic null distribution of the V_r is $N(0, 1)$ and that of $X_P^2 - V_1^2 - \ldots - V_r^2$ is χ^2_{m-1-r}. As long as the residual is significantly large, additional V_r should be calculated. Use of residuals in a similar way was recommended by Durbin and Knott (1972) with regard to the Cramer von Mises statistic. The distribution of the residual in that case is not a standard distribution, and access to tables may be a problem for some users. In Table 5.4 exact sizes have been calculated corresponding to 5% critical points of χ_1^2 for V_1^2 and V_2^2, χ^2_{m-2} for $X_P^2 - V_1^2$ and χ^2_{m-3} for $X_P^2 - V_1^2 - V_2^2$. These sizes agree well with the nominal 5%. Further calculations supporting this agreement may be found in Best and Rayner (1985a).

We will now examine a small power study to compare the performances

Table 5.4 Actual test sizes when the nominal size is 0.05

(m, n)	V_1^2	V_2^2	$V_1^2 + V_2^2$	X_P^2	$X_P^2 - V_1^2$	$X_P^2 - V_1^2 - V_2^2$
(5, 5)	0.0358	0.0266	0.0394	0.0366	0.0368	0.0394
(5, 10)	0.0559	0.0452	0.0415	0.0398	0.0396	0.0468
(5, 15)	0.0543	0.0529	0.0434	0.0414	0.0492	0.0454
(5, 20)	0.0473	0.0515	0.0501	0.0505	0.0503	0.0479
(10, 10)	0.0530	0.0480	0.0436	0.0395	0.0442	0.0394
(10, 15)	0.0472	0.0495	0.0462	0.0450	0.0443	0.0456

of V_1^2, V_2^2, $V_1^2+V_2^2$ and X_P^2. In this study, we will include the discrete analogue of the Kolmogorov–Smirnov statistic advocated, for example, by Conover (1980, p. 346), Horn (1977), and Pettitt and Stephens (1977). The statistic is

$$S = \max_{1\le r\le m} \left| \sum_{i=1}^{r} (N_i - np_i) \right|$$

Take H_0: $\pi_i = 1/m$ and the alternatives for $n = 12$ and $m = 6$ to be

$A_1(\delta)$: $\pi_i = \{i^\delta - (i-1)^\delta\}m^{-\delta}$

$A_2(\delta)$: $\pi_i = m^{-1} - \delta$, $i \le m/2$ and $\pi_i = m^{-1} + \delta$, $i > m/2$

$A_3(\delta)$: $\pi_i = \{1 - (i-2)\delta\}/6$, $i \le m/2$ and $\pi_i = \{1 + (i-5)\delta\}/6$, $i > m/2$.

The first two alternatives are trend alternatives with $n\pi_1 \le n\pi_2 \le \ldots \le n\pi_m$ or $n\pi_1 \ge n\pi_2 \ge \ldots \ge n\pi_m$ while the A_3 alternatives involve a decrease and increase in $n\pi_i$ values. Except for the A_3 alternative the S powers shown in our Table 5.5 are taken from Table 3 of Pettitt and Stephens (1977).

We note in Table 5.5 the parallel behavior of V_1^2 and S; both are good for the trend alternatives but neither is as good as $(V_1^2+V_2^2)$ for the A_3 alternatives. It appears then that S is not really a good omnibus test. This is in agreement with results obtained by Stephens (1974a) for the continuous counterpart of S, the Kolmogorov–Smirnov D statistic, where it was shown that D was competitive for detecting shifts in mean, but not in variance. See Table 3 of Stephens (1974a) and the related discussion.

Table 5.5 Exact powers of tests based on S, X_P^2, V_1^2, V_2^2, and $V_1^2+V_2^2$ with $n = 12$, $m = 6$, and $\alpha = 0.05$ for various A_i alternatives

Alternative	S	X_P^2	V_1^2	V_2^2	$V_1^2+V_2^2$
$A_1(.25)$	0.895	0.887	0.923	0.719	0.939
$A_1(.5)$	0.414	0.359	0.482	0.229	0.483
$A_1(.8)$	0.092	0.074	0.105	0.066	0.102
$A_1(1)$	0.050	0.050	0.050	0.050	0.050
$A_1(1.25)$	0.081	0.068	0.091	0.050	0.072
$A_1(1.5)$	0.163	0.110	0.203	0.050	0.143
$A_1(2)$	0.408	0.207	0.510	0.050	0.270
$A_1(3)$	0.815	0.627	0.927	0.120	0.836
$A_2(.033)$	0.089	0.070	0.090	0.050	0.076
$A_2(.066)$	0.223	0.139	0.219	0.050	0.165
$A_2(.1)$	0.476	0.290	0.458	0.050	0.354
$A_3(.5)$	0.140	0.144	0.093	0.289	0.263
$A_3(.65)$	0.194	0.222	0.105	0.449	0.398
$A_3(.8)$	0.265	0.333	0.118	0.629	0.561

By analogy with its continuous counterpart, it is likely that the reverse will be the case for the discrete Watson's U^2 statistic discussed by Freedman (1981). The discrete U^2 statistic will pick variance changes well, but will be less competitive for detection of mean shifts. Due to the weighting of its components, we suggest the statistic proposed by Hirotsu (1986) is likely to perform like S.

We suggest V_2 is sensitive to changes in variance. Its performance for the A_3 alternatives is superior, while its power is low for the trend alternatives where there is little variance shift to detect.

To summarize, examination of X_P^2 and its first two components provides a check on many alternatives. The first component, V_1, is good at detecting trend alternatives, as is S; the second component, V_2, is sensitive to variance changes; finally, $\{X_P^2 - V_1^2 - V_2^2\}$ provides a check on more complex alternatives.

We now turn to a comparison with Read's (1984a) statistics, which were discussed §2.3. Read (1984a) suggested H_0 should be tested using the statistics

$$PD(\lambda) = 2\{\lambda(\lambda+1)\}^{-1} \sum_{j=1}^{m} N_j[(N_j/(np_j))^{\lambda} - 1], \qquad -\infty < \lambda < \infty$$

Putting $\lambda = 1$ gives X_P^2, $\lambda = -0.5$ gives the Freeman–Tukey statistic, and $\lambda = 0$ gives the log-likelihood ratio statistic. Read suggested that $1/3 \le \lambda \le 2/3$ is a good choice for λ. His Table 1 showed that λ in this range gives statistics with power performance close to X_P^2. Following Read, consider alternatives

$$H_1(\delta): \pi_j = \{1 - \delta/(m-1)\}/m, \; j = 1, \ldots, m-1 \text{ and } \pi_m = (1+\delta)/m$$

Our Table 5.6 compares powers for $PD(\lambda)$ in which $\lambda = -.5$, .0, .5, and 1.0, taken from Read's Table 1, with exact powers of V_1^2 and $(V_1^2 + V_2^2)$. It

Table 5.6 Exact powers of tests based on $PD(\lambda)$, V_1^2, V_2^2, $(V_1^2 + V_2^2)$, and S when $\alpha = .05$, $n = 20$, $m = 5$ for various H_1 alternatives

	Alternative		
Statistic	$\delta = 1.5$	$\delta = 0.5$	$\delta = -0.9$
$PD(-0.5)$	0.336	0.080	0.588
$PD(0.0)$	0.610	0.107	0.447
$PD(0.5)$	0.681	0.119	0.322
$PD(1.0)$	0.700	0.123	0.272
V_1^2	0.655	0.134	0.252
V_2^2	0.528	0.108	0.211
$V_1^2 + V_2^2$	0.760	0.159	0.317
S	0.511	0.137	0.178

appears that $(V_1^2 + V_2^2)$ is competitive with the $PD(.5)$ statistic for the three alternatives shown.

5.4 Examples

From the power comparisons it seems that V_1^2 will reasonably detect shifts in mean, V_2^2 shifts in variance, and $V_1^2 + V_2^2$ will detect shifts in both mean and variance. Clearly $X_P^2 - V_1^2$, $X_P^2 - V_2^2$, and $X_P^2 - V_1^2 - V_2^2$ will detect residual variation apart from shifts in mean, variance, and both mean and variance, respectively. In a preliminary analysis, all these statistics could be calculated. In general, some compensation should be made when applying more than one test of significance. For the following examples, it may be assumed that the given component statistics of X_P^2 were known a priori to be appropriate. This can be confirmed by checking for residual variation. Calculating formulae for the components for nonequiprobable cases are given in Appendix 3.

We hope the reader will see from the examples that V_1 rather than X_P^2 is the appropriate competitor for S, since both reasonably detect mean shifts. The components of X_P^2 complement each other nicely, however, detecting other alternatives as well as residual variation.

Example 5.4.1

Suppose, as in Pettitt and Stephens (1977) we have $(n_1, n_2, n_3, n_4, n_5) = (0, 1, 0, 5, 4)$ and H_0: $p = .2, i = 1, \ldots, 5$. Simple calculations give $S = 5$ and $X_P^2 = 11$. From Table 1 of Pettitt and Stephens (1977) we find $P(S \geq 5) = .00477$ while the usual χ^2 approximation gives $P(X_P^2 \geq 11) = .024$. The exact probability is .037. Thus S appears more sensitive than X_P^2. Using the formulae in Appendix 3 or otherwise,

$$V_1 = \{\Sigma_j (j-1)N_j - 20\}\sqrt{20} = 2.6833$$

and using the χ_1^2 approximation, $P(V_1^2 \geq 7.2) = .007$. Results in Best and Rayner (1985a) indicate that χ_1^2 approximation will be excellent in this equiprobable case, and indeed the exact value is .008.

Example 5.4.2

Horn (1977) considered an example in which $(n_1, n_2, n_3, n_4, n_5, n_6) = (0, 15, 4, 7, 2, 2)$ and in which $(np_1, np_2, np_3, np_4, np_5, np_6) = (1, 17, 7, 3, 1, 1)$. $X_P^2 = 9.854$ and $P(X_P^2 \geq 9.854) = .09$ using the χ_5^2 approximation, while an exact calculation gives 0.0875.

$$S^- = \sup_{1<r<m} \left\{ \sum_{j=1}^{r} (np_j - N_j) \right\} = 6$$

and using $P(S^- \geq 6) = .5P(S \geq 6)$ and the unequal np_j approximation of 3.2(c) of Pettitt and Stephens (1977), we find, on using these authors' Table

1, that $P(S^- \geq 6) = .04$. Horn (1977) gives an exact value of .026. Again, it appears that S is more sensitive than X_P^2. However, $V_1 = 2.264$ and $P(V_1 \geq 2.264) = .011$ using the normal approximation. The exact probability is .019.

Example 5.4.3

Let us permute the observed counts in Example 5.4.1 to obtain $(n_1, n_2, n_3, n_4, n_5) = (4, 0, 1, 0, 5)$. Then X_P^2 remains unaltered, but $S = 3$ and $P(S \geq 3) > .042$ and so S is now less sensitive than X_P^2. As there is no natural order in the permutations, the use of S and V_1 is perhaps not warranted here.

Example 5.4.4

Data on the incidence of a disease are often given as monthly totals, and it is often important to test for absence of seasonal variation in the incidence counts. For example, Edwards (1961) gives the monthly counts $(10, 19, 18, 15, 11, 13, 7, 10, 13, 23, 15, 22)$ for first-born anencephalics from Birmingham during 1940–1947. If we take the null expectation for each month as $176/12 = 14.6667$, we find $S = 16$ and $X_P^2 = 18.73$. For both statistics, $P > .05$. However, $V_2 = 2.06$, which gives $P < .05$ for a 2-sided test. This is in agreement with the value of 6.4 found from Edwards' test. The seasonal variation evident in Figure 5.1 is confirmed.

Example 5.4.5

Suppose n consumers or judges rank k by products using the integers $1, 2, \ldots, k$. A conventional method of comparing the k products is to calculate Friedman's χ_r^2 statistic or, equivalently, Kendall's concordance coefficient, W. If counts are made of each of the $k!$ possible rankings by a consumer, however, then an alternative analysis would be to calculate X_P^2 and S, as earlier, with $m = k!$, $p_j = m^{-1}$, $j = 1, \ldots, m$. In a consumer preference study with, for example, $k \leq 4$, and $n \geq m$, there would not be many zero counts.

Consider the following 18 rankings of 3 products, where ijk means product 1 is given rank i, product 2 is given rank j, and product 3 is given rank k:

$$132, 231, 231, 321, 321, 321, 231, 321, 231,$$
$$231, 312, 312, 132, 321, 231, 312, 123, 132.$$

For the m possible permutations $(123, 132, 213, 231, 312, 321)$, the counts are $(n_1, n_2, n_3, n_4, n_5, n_6) = (1, 3, 0, 6, 3, 5)$, and so $X_P^2 = 8.67$, $V_1 = 1.79$, and $S = 5.0$; thus, $P(X_P^2 \geq 8.67) = .13$, $P(S \geq 5) = .064$, and $P(V_1^2 \geq 3.20) = .078$. These last two probabilities indicate possible differences between the counts.

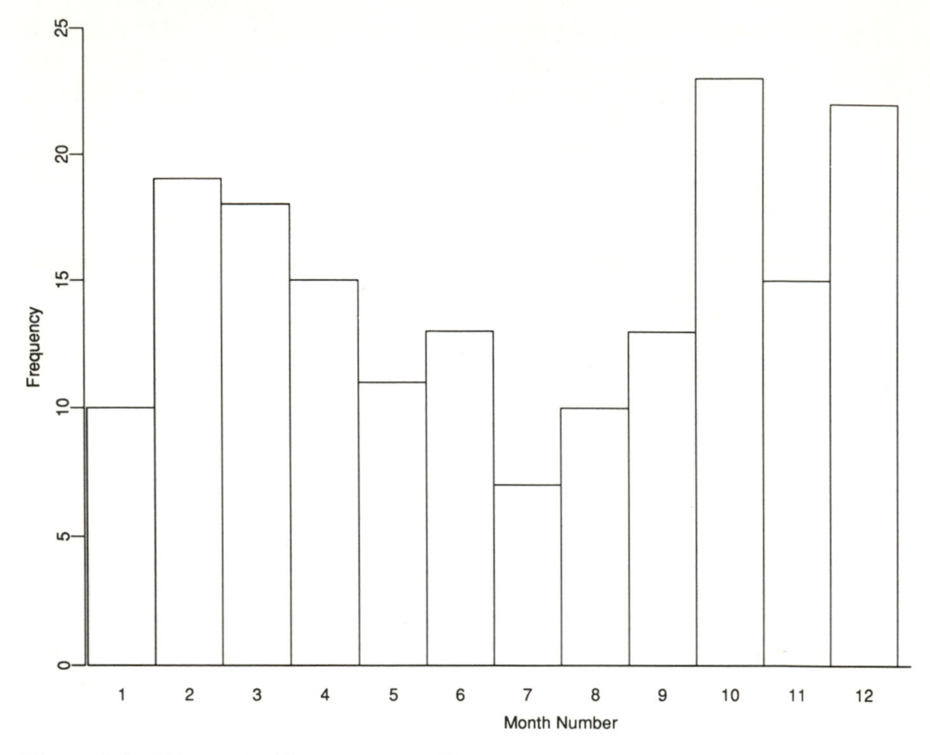

Figure 5.1 Disease incidence per month.

We will now show how *other* orthonormal functions may be appropriate. In our present example, suppose we wished to compare products. Let R_p be the sum of the ranks for product p, with $p = 1$, 2, and 3. In our example $R_1 = 1 + 2 + 2 + \ldots = 40$, $R_2 = 3 + 3 + 3 + \ldots = 42$, and $R_3 = 2 + 1 + 1 + \ldots = 26$. These sums may also be obtained from the counts N_i, $i = 1, 2, \ldots, 6$, by using

$$R_1 = N_1 + N_2 + 2(N_3 + N_4) + 3(N_5 + N_6)$$

$$R_2 = N_3 + N_5 + 2(N_1 + N_6) + 3(N_2 + N_4)$$

$$R_3 = N_4 + N_6 + 2(N_2 + N_5) + 3(N_1 + N_3)$$

We can partition X_P^2 into single degrees of freedom a priori contrasts. Here these contrasts might be

$$R_1 - R_2 = -N_1 - 2N_2 + N_3 - N_4 + 2N_5 + N_6,$$

and

$$(R_1 + R_2)/2 - R_3 = 3(-N_1 - N_3 + N_4 + N_6)/2$$

We wish to define V_r^* involving these contrasts, but satisfying $V^* = (V_r^*) = HN/\sqrt{n}$, where H satisfies Equation 5.1.8. This is achieved if we define $V_r^* = \Sigma_{j=1}^{6} l_{rj} N_j / \sqrt{(np_j)}$ where $\sqrt{(np_j)} = \sqrt{3}$ and for $j = 1, \ldots, 6$, $\sqrt{(12)} l_{1j} =$

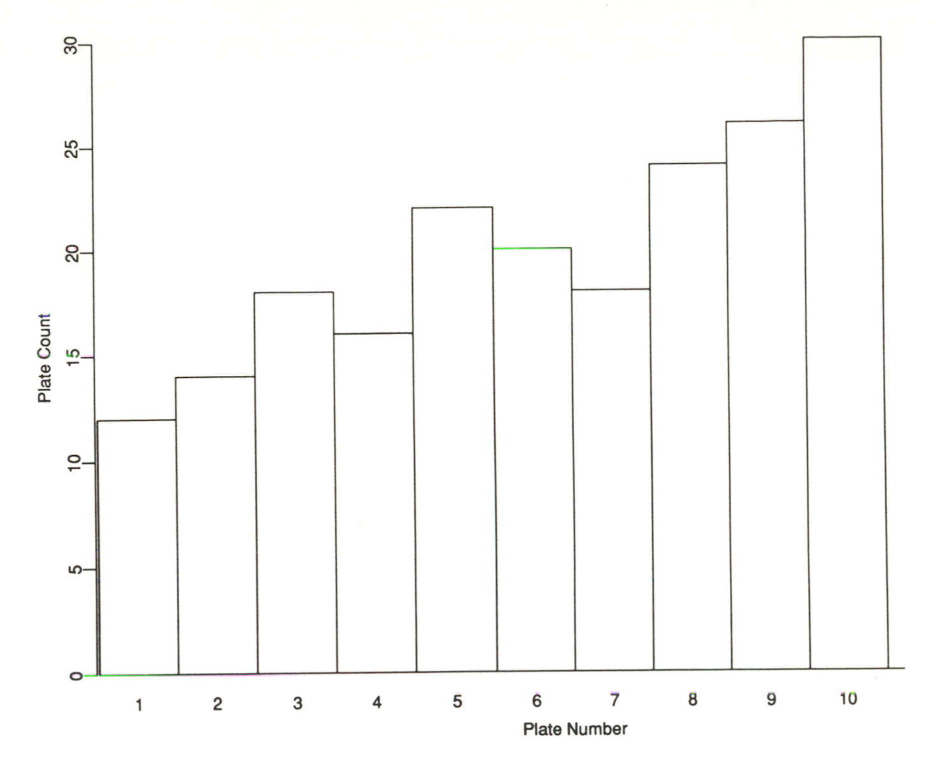

Figure 5.2 Termite counts.

−1, −2, 1, −1, 2, 1, and $2l_{2j} = -1, 0, -1, 1, 0, 1$, respectively. Then l_{rj}, $r = 3, 4, 5$ are constructed to satisfy (l_{rs}) orthogonal. With these definitions $V_1^* = (R_1 - R_2)/6$ and $V_2^* = \{(R_1 + R_2)/2 - R_3\}/\{3\sqrt{3}\}$, and are obviously standardizations of the required contrasts. As usual, $X_P^2 = (V_1^*)^2 + \ldots + (V_5^*)^2$. Now $(V_1^*)^2 + (V_2^*)^2$ is equal to Friedman's χ_r^2, and is a partition of X_P^2, as was shown by Anderson (1959).

In our example, $(V_1^*)^2 = 0.111$ and $(V_2^*)^2 = 8.333$. The nonsignificance of V_1^* indicates products 1 and 2 received similar ranks, whereas the significance of V_2^* indicates that product 3 was ranked better than the average of the ranks of 1 and 2. We suggest calculation of X_P^2 and its components V_1^* and V_2^* gives a more informative analysis than χ_r^2 for this data set.

Example 5.4.6

Chacko (1966) proposed a test for ordered alternatives to the equiprobable multinomial for which he quoted the following example. A plate with the humidity values continuously decreasing was divided into 10 equal parts and 20 termites introduced on each part (Figure 5.2). The number of termites counted after a specified time interval on each of the 10 parts of the plate

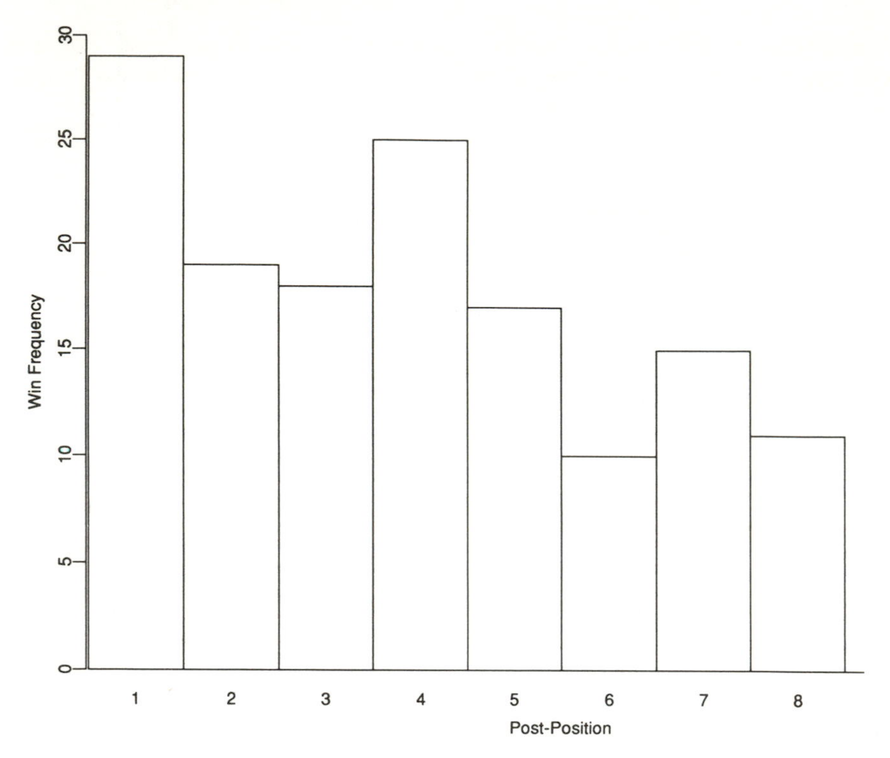

Figure 5.3 Horse race winners and post-position.

were

$$12, 14, 18, 16, 22, 20, 18, 24, 26, 30$$

Calculation of X_P^2 gives a value of 14.00, which, on 9 degrees of freedom, gives a P value in excess of 10%. Chacko's test gives $P = .0023$.

Now ordered alternatives are well modeled by a mean shift, such as would be given by $h_1(x)$ linear in §4.2. The corresponding alternative for discrete alternatives is to use the linear discrete orthogonal polynomial in §5.3. Before sighting the data, therefore, a test based on the first component V_1 is a sensible test to apply. If we do so, we obtain $V_1 = 3.45$, and V_1^2/X_P^2 is approximately 85%. Using the χ^2 approximation, as is reasonable for $n = 200$ and equal cell probabilities, we obtain $P(V_1^2 \geq 11.88) < .001$. Thus, the test based on V_1^2 is even more critical of the data than is Chacko's test.

Example 5.4.7

Lee (1987) considered some horse-racing data. Punters often maintain that in a race around a circular track, horses whose starting positions are nearest the rail have an advantage. Table 1 of Lee (1987) listed the post position (1 is the inside position, nearest the rail, and 8 is the outside position, furthest from the rail) and the number of winners from each post position in 144 races.

Post position	1	2	3	4	5	6	7	8
Number of winners	29	19	18	25	17	10	15	11

Under the null hypothesis that starting position does not affect the number of winners, we would expect 18 winners from each position. A value of 16.33 for X_P^2 results, yielding a P value of .02 using the usual χ_7^2 approximation. Thus, the null hypothesis is discredited. Figure 5.3 suggests a trend with the number of winners decreasing with distance from the rail. To check this, we calculate $\hat{V}_1 = -3.24$, $\hat{V}_1^2/X_P^2 = 0.64$ and $P(\hat{V}_1 < -3.24) =$ .0005 (approximately), thereby confirming the impression from Figure 5.3. We suggest our analysis is as informative as that given by Lee (1987).

5.5 A More Comprehensive Class of Tests

In this section, a modification is made of the model examined in §5.1. The aim is to introduce some flexibility in the models we are examining; this flexibility is exploited in §5.6. Inasmuch as this section is mainly concerned with the derivation of a score statistic, some readers may wish to skip or skim this section. Both §5.5 and §5.6 could be omitted on first reading of this monograph without loss of continuity.

Recall that in §5.1 we took as an alternative to the null hypothesis that the m cell probabilities are

$$\pi_j = C(\theta)\exp\left\{\sum_{i=1}^{k}\theta_i h_{ij}\right\}p_j, \qquad j = 1, \ldots, m \tag{5.5.1}$$

Now suppose B is a $k \times s$ matrix of constants of rank s, and we wish to test the null hypothesis $\phi = (\phi_r) = 0$, where ϕ is an $s \times 1$ vector of parameters defined by $\theta = B\phi$. It will be shown that, with this formulation, orthogonal components of the test statistic may be easily obtained and the corresponding parameters, which each component optimally detects, may be easily identified.

As before, a random sample of n observations is taken and N_j, the number of observations in the jth cell, $j = 1, \ldots, m$, is noted. Write n_j for a realization of N_j, and note that $n = \sum_j n_j = \sum_j N_j$. Now the logarithm of the likelihood is

$$l = \text{constant} + \sum_{j=1}^{m} n_j \ln \pi_j$$

so that

$$\frac{\partial l}{\partial \phi_r} = \sum_{i,j}\frac{n_j}{\pi_j}\frac{\partial \pi_j}{\partial \theta_i}\frac{\partial \theta_i}{\partial \phi_r} = \sum_{i,j} n_j\{h_{ij} - E_k(H_i)\}b_{ir}$$

Recall from Equation 5.1.4 that $E_k(H_i) = \sum_j h_{ij}\pi_j = -\partial \log C(\theta)/\partial\theta_i$. Put $N = (N_j)$, $H = (h_{ij})$, and $p = (p_j)$, and note that under H_0, $E_0(H_i) = \sum_j h_{ij} p_j = (Hp)_i$. The efficient score is

$$U = \left(\frac{\partial l}{\partial \phi_r}\right)\bigg|_{\phi=0} = B^T HN - nB^T Hp$$

As in the derivation of Theorem 5.1.1, $\partial E_k(H_i)/\partial\theta_u = \text{cov}_k(H_i, H_u)$. It follows that

$$\frac{\partial^2 l}{\partial\phi_r\,\partial\phi_s} = -\sum_{i,j,u} n_j \frac{\partial E_k(H_i)}{\partial\theta_u}\frac{\partial\theta_u}{\partial\phi_s} b_{ir}$$

$$= -n\sum_{i,u} \text{cov}_k(H_i, H_u) b_{us} b_{jr}$$

The information matrix is thus $I = nB^T\Sigma B$, where Σ is the $k \times k$ matrix with rsth element $\sigma_{rs} = \text{cov}_0(H_r, H_s)$. It follows that $\Sigma = HC_V H^T$, where $C_V = \text{diag}(p_j) - pp^T$ is the covariance matrix of N. We now have that the score statistic is

$$S = U^T I^{-1} U = (N - np)^T H^T B[B^T H C_V H^T B]^{-1} B^T H(N - np)/n \quad (5.5.2)$$

This has asymptotic distribution χ^2_s, central under the null hypothesis $\phi = 0$, and noncentral under contiguous alternatives of the form $\phi = (a_i)/\sqrt{n}$ for constants $a_1, \ldots, a_s$.

If Σ is of full rank k, we may take $s = k$, $B = \Sigma^{-1/2}$ so that the inverse required in Equation 5.5.2 is that of the unit matrix. Now define $V = (V_i) = \Sigma^{-1/2} H(N - Np)/\sqrt{n}$ so that $S = S_k$ (say) $= V_1^2 + \ldots + V_k^2$. The V_i constitute an orthogonal decomposition of S in the sense that each V_i detects ϕ_i and no other ϕ_j because by starting with Equation 5.5.1, and taking B as the ith row of $\Sigma^{-1/2}$ we may derive V_i^2 as the score statistic for testing the null hypothesis $\phi_i = 0$. It is therefore weakly optimal for detecting ϕ_i and asymptotically has a χ^2_1 distribution, central under $\phi_i = 0$ and noncentral under contiguous alternatives of the form $\phi_i(n) = a_i/\sqrt{n}$. Since V may be shown to have mean 0 and unit covariance matrix, and by the Central Limit Theorem is asymptotically m-variate normal, the V_i are asymptotically independent. So S is the sum of m asymptotically independent χ^2_1 random variables.

Example 5.5.1

For the equiprobable case $p_j = 1/m$, $j = 1, \ldots, m$, Seyb (1984) investigated the test that results from choosing

$$h_{ju} = (u^{j+1} - u^j)/\{(j+1)m^{j+1}\}$$

This is equivalent to using powers in the corresponding continuous model, which is similar, for example, to Kopecky and Pierce (1979). The resulting test performs quite well, but is unnecessarily complicated.

The inclusion of B in the Equation 5.5.1 enables asymptotically orthogonal components to be constructed, whatever the choice of H. In the next section, we will exploit the choices given by having a more comprehensive model.

5.6 Overlapping Cells Tests

The main advantage that the tests we are about to construct have over Pearson's X^2 test is that the cells may be amalgamated to improve the χ^2 approximation, without necessarily losing the information in any cell. The cost, compared with Pearson's test, is a more complicated test statistic. A competitor test was given by Hall (1985); his statistic is itself simple, but it has a complicated distribution.

It should be emphasized that a family of tests is derived. The user has the flexibility to construct statistics that are appropriate for the problem at hand. This would usually be done to improve, relative to Pearson's X^2 test, either the null χ^2 approximation or the power against specified alternatives.

Best and Rayner (1985a) and Rayner, Best, and Dodds (1985) suggested that low-order alternatives, i.e., those with m "small" in Equation 5.1.1, are of some practical importance. Taking too many classes in Pearson's X^2 test gives reasonable power against high-order alternatives, and little information is lost due to categorization, but high-order alternatives are not often important. Taking too few classes gives good detection of low-order alternatives, but loses information by overcategorization. Keeping the number of classes large but looking only at the first, two, three, or four components, for example, gives a reasonable compromise in the detection of low order alternatives. The power studies in Best and Rayner (1985a, 1987a) confirm this.

In the following, the formulation of §5.5 is used to produce overlapping cells X^2 tests. Since these tests are based on score statistics, they are optimal for detecting specified alternatives and have test statistics with convenient null and alternative distributions. The penalty is that a matrix inverse is required in the calculation of the statistics. The overlapping cells statistics have the advantage that if cell expectations are small, combining cells in this manner can give a moderate expectation for the combined cells, and the χ^2 approximation to the null distribution will be more accurate.

In view of the preceding comments, a reasonable strategy for the detection of low-order alternatives would appear to be to keep m, the number of categories, as large as is feasible to avoid loss of information by overcategorization, and to overlap cells until all the combined cell expectations are moderate, i.e., at least 5. Next, calculate a test statistic based on three components, using $S_3 = V_1^2 + V_2^2 + V_3^2$, for example, to best detect the low-order alternatives. The only obvious difficulty with this approach is that the alternatives the V_i detect may not be readily interpretable because B must be chosen to be $\Sigma^{-1/2}$.

Hall (1985) defined an overlapping cells X^2 test by grouping successive batches of r adjacent cells. His test has desirable properties as both n and k approach infinity, but the asymptotic distribution of the test statistic is inconvenient for practical use. He considered only cells equiprobable under the null hypothesis, and the matrix in the quadratic form defining his test statistic is the "wrong" $\text{diag}(r^{-1}, \ldots, r^{-1})$ instead of $(HVH^T)^{-1}$, with $B = I_m$ in Equation 5.5.2. We now use the "right" matrix—the asymptotic covariance matrix—in defining the quadratic form. Equiprobable cells will not be assumed, and this affects which cells should be overlapped.

In Equation 5.5.2 define

$$h_{ij} = 1 \text{ for } j = i + 1 \ (\text{mod } r), \ldots, i + r(\text{mod } r),$$
$$= 0 \text{ otherwise.} \tag{5.6.1}$$

It may be shown that this test, which takes successive groups of r adjacent cells, is equivalent to that formed by taking successive groups of the complementary $m - r$ adjacent cells. In particular, the test with $r = m - 1$ is the same as the test based on X_P^2. So there is no point in taking $r > [m/2]$, the integer part of $m/2$.

The main difficulty with the tests using Equation 5.6.1 is that they are unnecessarily restrictive. If the successive cell probabilities are, .5, .3, .1, .05, .03, .02, for example, then it is the last three cells that need to be overlapped with cells with high expectation, so that the combined cell expectations are increased. Routinely overlapping the first two and the second two cells is unnecessary to obtain reasonable combined cell expectations. A better approach would be to combine the first cell with each of the remaining cells by taking

$$h_{ij} = 1 \text{ for } j = 1 \text{ and } j = i + 1, \qquad i = 1, \ldots, 5 \tag{5.6.2}$$

Examples similar to this will be given subsequently. Sizes are simulated to verify that combining cell expectations improves on not combining cell expectations with regard to the closeness of the exact and nominal sizes (obtained using χ^2 percentage points).

Of course, there are many possible choices of H. These correspond to different tests, and we must ask which of these are to be preferred? Exact size can hardly be the criterion, because if the cell expectations are large enough (i.e., not less than 5), the exact and nominal sizes should be very close. Power is unlikely to be the answer. Being score tests, they are all asymptotically optimal, although against different alternatives. The alternatives would seem to be the key. If possible, the test should be chosen so that interpretable and meaningful alternatives can be detected.

Finally, a word about components (which were defined in the previous section). Each component V_i corresponds to a different test, and again, whether to use them or not is a difficult decision. It is valuable to be able to dissect the adequacy of the model fit, and we can write down the alternative each component optimally detects. Some ingenuity, however, may be required in interpreting these alternatives. We will subsequently assume this question is settled, and only the closeness of the actual and nominal sizes will be examined in a small simulation study.

For $m = 5$ and $n = 10$ and cell probabilities of

(a) .01, .01, .01, .01, and 0.96, and
(b) .05, .05, .05, .05, and .80,

the sizes of the tests corresponding to three different H's and their orthogonal components were approximated using 5,000 simulations. The H's were specified by

1. $(H_1)_{ij} = 1$ for $j = i + 1$ and $j = i + 2, i = 1, \ldots, 4,$ and $(H_1)_{ij} = 0$ otherwise;
2. $(H_2)_{ij} = 1$ for $j = i$ and $j = 5, i = 1, \ldots, 4,$ and $(H_2)_{ij} = 0$ otherwise; and
3. $(H_3)_{1j} = \sqrt{0.2}, j = 1, \ldots, 5,$
   $(H_3)_{ij} = 1/\sqrt{[i(i-1)]}, j = 1, \ldots, i - 1,$ and
   $(H_3)_{ii} = -\sqrt{[(i - 1/i]}, i = 2, 3, 4.$

H_1 routinely combines pairs of adjacent cells, and corresponds to Hall's approach. H_2 combines each small probability cell with the high probability cell. H_3 combines cells only in the sense that it contrasts them. Appropriately augmented, it is a standard Helmertian matrix (see Lancaster, 1965, for information on the Helmert matrices). If the test has m degrees of freedom, it makes comparisons between the first j cells and the $(j + 1)$th cell, for $j = 1, \ldots, m$. This comparison effectively combines the first $j + 1$ cells. The results of the simulation are given in Table 5.7. Each tabulated value has a standard error of approximately .003.

The results are of course most extreme in case (a), where the smallest cell expectations are quite small. The H_3 sizes are extremely variable and of the three, the most removed from 5%, averaging 11%; the H_1 sizes are more

Table 5.7 Simulated sizes multiplied by 1,000, with $n = 10$, $k = 5$, and nominal size 5%; based on 5,000 simulations

	V_1^2	V_2^2	V_3^2	V_4^2	S_2	S_3	S_4
	(a) *Cell probabilities* .01, .01, .01, .01, *and* .96						
H_1	27	95	95	15	102	186	60
H_2	37	39	36	37	52	58	59
H_3	173	98	81	60	261	58	60
	(b) *Cell probabilities* .05, .05, .05, .05, *and* .80						
H_1	69	36	52	66	46	57	73
H_2	67	72	71	67	41	63	76
H_3	120	51	64	33	33	59	71

reasonable, averaging 8%, but with some extreme values; the H_2 sizes are uniformly acceptable, averaging 4.5%. In case (b) the cell expectations are larger and there is little to choose from between the tests based on the different H_i. The performance of V_1^2, however, is still not satisfactory. If it were, then X_P^2 and its components based on H Helmert would be the preferred test statistic unless special comparisons were required.

The test statistics assessed were the components V_1^2, V_2^2, V_3^2, V_4^2 and the partial sums $S_s = V_1^2 + \ldots + V_s^2$, $s = 2, 3, 4$.

The tests considered will perhaps be most useful for very sparse multinomials and for extremely unbalanced nonequiprobable situations. The suggestion is that the overlapping cells X^2 tests be employed to increase cell expectations and thus accelerate the approach to the asymptotic distribution and the attainment of the asymptotic properties. The method is virtually pooling, but without the loss of degrees of freedom and, therefore, power.

6

Neyman Smooth Tests for Uncategorized Composite Null Hypotheses

6.1 Neyman Smooth Tests for Uncategorized Composite Null Hypotheses

Suppose we wish to test the null hypothesis that $X_1, \ldots, X_n$ is a random sample from a continuous distribution with probability density function $f(x; \beta)$, where $\beta = (\beta_1, \ldots, \beta_q)^T$ is a vector of nuisance parameters. The first step is to imbed the null probability density function in an order k alternative,

$$g_k(x; \theta, \beta) = C(\theta, \beta) \exp\left\{\sum_{i=1}^{k} \theta_i h_i(x, \beta)\right\} f(x; \beta) \tag{6.1.1}$$

where $\{h_i(x; \beta)\}$ are complete and orthonormal on $f(x; \beta)$ with $h_0(x; \beta) \equiv 1$, and where we assume the existence of a normalizing constant $C(\theta, \beta)$. Testing for $f(x; \beta)$ is equivalent to testing $H_0: \theta = (\theta_1, \ldots, \theta_k)^T = 0$ against $K: \theta \neq 0$.

Subsequently, we will require $f(x; \beta)$ to be "regular." Assume we are given a simple random sample from the distribution with probability density function $g_k(x; \theta, \beta)$. By regular we mean that derivatives of the logarithm of the likelihood with respect to elements of θ and β up to second order exist, as do their expectations. Furthermore, we assume that the integral of Equation 6.1.1 over the reals may be differentiated under the integral; such differentiations with respect to the θ_i are permitted by Lehmann (1986, p. 59). In specific cases, the differentiability with respect to the β_j must be checked.

The score statistics are derived for regular cases of Equation 6.1.1. This confers the weak optimality enjoyed by score tests upon the smooth tests that result, and ensures that the test statistics have asymptotic χ^2 distributions under the null hypothesis and under contiguous alternatives of the form $(a_1, \ldots, a_k)^T/\sqrt{n}$, where $a_1, \ldots, a_k$ are constants. Furthermore, the choice of orthonormal functions in the definition of the order k alternative ultimately gives a computationally convenient statistic with easily accessible components. Kopecky and Pierce (1979) and Thomas and Pierce (1979) used powers, $h_i(x;\beta) = F^i(x;\beta)$, where $F(x;\beta)$ is the cumulative distribution function corresponding to $f(x;\beta)$. Their tests involve quadratic forms and so require tables of coefficients to implement; ours only involve sums of squares. Additionally, their components are not as easily interpreted as those we give. The first component of our tests is often a well-known goodness of fit test in its own right.

We take the order k probability density function to be Equation 6.1.1. Now note that θ and $h = (h_i(x;\beta))$ are $k \times 1$, and take $\{h_i(x;\beta)\}$ to be a complete orthonormal set on $f(x;\beta)$, so that

$$E_0[h_r(X;\beta)h_s(X;\beta)] = \delta_{rs}$$

The zero subscripted expectation is taken with respect to the null probability density function $f(x;\beta)$. In particular, we take $h_0(x;\beta) \equiv 1$ so that $E_0[h_r(X;\beta)] = \delta_{ro} = 0$ for $r = 1, 2, 3, \ldots$. The operator E_k denotes expectation with respect to the order k probability density function $g_k(x;\beta)$. All expectations are subsequently assumed to exist.

To derive score statistics, we start with the likelihood that, given a random sample $x_1, \ldots, x_n$ has logarithm

$$l = n \log C + \sum_{i,j} \theta_i h_{ij} + \sum_j \log f_j$$

where $C = C(\theta;\beta)$, $h_{ij} = h_i(x_j;\beta)$ and $f_j = f(x_j;\beta)$. The derivatives of l involve derivatives of $\log C$ that we will now derive.

Lemma 6.1.1. The derivatives of $\log C$ with respect to θ_r and β_u are given by

$$\frac{\partial \log C}{\partial \theta_r} = -E_k[h_r] \tag{6.1.2}$$

and

$$\frac{\partial \log C}{\partial \beta_u} = -\sum_i \theta_i E_k\left[\frac{\partial h_i}{\partial \beta_u}\right] - E_k\left[\frac{\partial \log f}{\partial \beta_u}\right] \tag{6.1.3}$$

in which $h_i \equiv h_i(x;\beta)$.

Proof. Differentiating

$$\int_{-\infty}^{\infty} g_k(x;\theta,\beta)\,dx = 1$$

with respect to θ_r gives Equation 6.1.2, and differentiating with respect to β_u gives Equation 6.1.3.

Differentiating l now gives

$$\frac{\partial l}{\partial \theta_r} = \sum_j h_{rj} + n\frac{\partial \log C}{\partial \theta_r} = \sum_j (h_{rj} - E_k[h_r])$$

on using Equation 6.1.2. On using Equation 6.1.3,

$$\frac{\partial l}{\partial \beta_u} = \sum_{i,j} \theta_i \frac{\partial h_{ij}}{\partial \beta_u} + \sum_j \frac{\partial \log f_j}{\partial \beta_u} + n \frac{\partial \log C}{\partial \beta_u}$$

$$= \sum_j \left\{ \frac{\partial \log f_j}{\partial \beta_u} - E_k\left[\frac{\partial \log f}{\partial \beta_u}\right]\right\} + \sum_{i,j} \theta_i \left\{ \frac{\partial h_{ij}}{\partial \beta_u} - E_k\left[\frac{\partial h_f}{\partial \beta_u}\right]\right\}$$

Subsequent differentiations will use the following result.

Lemma 6.1.2. Provided the quantities involved exist, if $D = D(X; \beta)$, then

$$\frac{\partial E_k[D]}{\partial \theta_t} = \text{cov}_k(h_t, D),$$

and

$$\frac{\partial E_k[D]}{\partial \beta_w} = E_k\left[\frac{\partial D}{\partial \beta_w}\right] + \sum_i \theta_i \,\text{cov}_k\left(D, \frac{\partial h_i}{\partial \beta_w}\right) + \text{cov}_k\left(\frac{\partial \log f}{\partial \beta_w}, D\right)$$

Proof.

$$\partial E_k[D]/\partial \theta_t = ((\partial C/\partial \theta_t)/C)E_k[D] + E_k[Dh_t]$$

$$= \text{cov}_k(h_t, D)$$

on using Equation 6.1.2, and

$$\partial E_k[D]/\partial \beta_w = E_k[\partial D/\partial \beta_w] + ((\partial C/\partial \beta_w)/C)E_k[D]$$

$$+ E_k\left[D \sum_i \theta_i(\partial h_i/\partial \beta_w)\right] + E_k[(\partial \log f/\partial \beta_w)D]$$

$$= E_k\left[\frac{\partial D}{\partial \beta_w}\right] + E_k\left[D \sum_i \theta_i \frac{\partial h_i}{\partial \beta_w}\right] + E_k\left[\frac{\partial \log f}{\partial \beta_w}\right]$$

$$- E_k[D]\left\{\sum_i \theta_i E_k\left[\frac{\partial h_i}{\partial \beta_w}\right] + E_k\left[\frac{\partial \log f}{\partial \beta_w}\right]\right\}$$

on using Equation (6.1.3). The expression in the lemma statement follows immediately.

The second order derivatives of l are thus given by

$$\frac{\partial^2 l}{\partial \theta_r \,\partial \theta_s} = n \frac{\partial \log C}{\partial \theta_r \,\partial \theta_s} = -n \,\text{cov}_k(h_r, h_s)$$

$$\frac{\partial^2 l}{\partial \theta_r \,\partial \theta_u} = -n \,\text{cov}_k\left\{h_r, \frac{\partial \log f}{\partial \beta_u}\right\} + \sum_j \left\{\frac{\partial h_{rj}}{\partial \beta_u} - E_k\left[\frac{\partial h_r}{\partial \beta_u}\right]\right\}$$

$$- n \sum_f \theta_f \,\text{cov}_k\left(h_r, \frac{\partial h_f}{\partial \beta_u}\right)$$

and

$$\frac{\partial^2 l}{\partial\beta_u \partial\beta_v} = \sum_j \frac{\partial^2 \log f}{\partial\beta_u\, \partial\beta_v} - n \operatorname{cov}_k\left(\frac{\partial \log f}{\partial\beta_u}, \frac{\partial \log f}{\partial\beta_v}\right) - nE_k\left[\frac{\partial^2 \log f}{\partial\beta_u\, \partial\beta_v}\right]$$

$$+ \sum_{j,f} \theta_f \frac{\partial}{\partial\beta_v}\left\{\frac{\partial h_{fj}}{\partial\beta_u} - E_k\left[\frac{\partial h_f}{\partial\beta_u}\right]\right\}$$

It follows that

$$- E_k\left[\frac{\partial^2 l}{\partial\theta_r\, \partial\beta_s}\right]\Bigg|_{\theta=0} = n\, \delta_{rs} = (I_{\theta\theta})_{rs}$$

using the orthonormality condition,

$$- E_k\left[\frac{\partial^2 l}{\partial\theta_r\, \partial\beta_u}\right]\Bigg|_{\theta=0} = n \operatorname{cov}_0\left(h_r, \frac{\partial \log f}{\partial\beta_u}\right) = (I_{\theta\beta})_{ru},$$

and

$$- E_k\left[\frac{\partial^2 l}{\partial\beta_u\, \partial\beta_v}\right]\Bigg|_{\theta=0} = n \operatorname{cov}_0\left(\frac{\partial \log f}{\partial\beta_u}, \frac{\partial \log f}{\partial\beta_v}\right) = (I_{\beta\beta})_{uv}$$

The score statistic has the form $S(\beta) = U_\theta^T \Sigma^{-1} U_\theta$ where $U_\theta = (h_f(X_1; \beta) + \ldots + h_f(X_n; \beta))$ and Σ is the asymptotic covariance matrix given by

$$\Sigma = I_{\theta\theta} - I_{\theta\beta} I_{\beta\beta}^{-1} I_{\beta\theta} = nM,$$

where

$$M = I_k - \operatorname{cov}_0\left(h, \frac{\partial \log f}{\partial\beta}\right)\left\{\operatorname{cov}_0\left(\frac{\partial \log f}{\partial\beta}, \frac{\partial \log f}{\partial\beta}\right)\right\}^{-1} \operatorname{cov}_0\left(\frac{\partial \log f}{\partial\beta}, h\right)$$

The usable form of the score test is $S(\hat{\beta})$, where $\hat{\beta}$ is the maximum likelihood estimator of β. Write $\hat{M}$ for M with β replaced by $\hat{\beta}$. The results so far can be summarized in a theorem.

Theorem 6.1.1. The score statistic for testing $\theta = 0$ with the regular model 6.1.1 is, provided $\hat{M}$ is nonsingular,

$$S(\hat{\beta}) = \left\{\sum_j h(X_j; \hat{\beta})/\sqrt{n}\right\}^T \hat{M}^{-1}\left\{\sum_j h(X_j; \hat{\beta})/\sqrt{n}\right\}$$

To avoid the problem of $\hat{M}$ being singular, and to enable components to be easily found, a modification of the theorem is made. Define φ by $\theta = B\varphi$, where B is a $k \times p$ matrix of elements $b_{ij} = b_{ij}(\beta)$ that depend on β; $\hat{B} = (b_{ij}(\hat{\beta}))$ should be chosen so that $\hat{B}^T\hat{M}\hat{B}$ is non-singular. Repetition of the approach that led to Theorem 6.1.1 gives Theorem 6.1.2.

Theorem 6.1.2. Provided $\hat{B}^T\hat{M}\hat{B}$ is nonsingular, the score statistic for testing $\varphi = 0$ with the regular model 6.1.1 is

$$S(\hat{\beta}) = \left\{\sum_j h(X_j; \hat{\beta})/\sqrt{n}\right\}^T \hat{B}(\hat{B}^T\hat{M}\hat{B})^{-1}\hat{B}^T\left\{\sum_j h(X_j; \hat{\beta})/\sqrt{n}\right\} \quad (6.1.4)$$

Statistic 6.1.4 can also be derived as the score statistic for testing $\theta = 0$ against $\theta \neq 0$ for the model

$$g_k(x; \theta, \beta) = C(\theta, \beta) \exp(\theta^T B^T h) f(x; \beta)$$

We are effectively testing for the projection of the null probability density function into the space spanned by $k - q$ of the $h_i(x; \beta)$. The alternative is defined in terms of k of the $h_i(x; \beta)$, but the q maximum likelihood equations for the β_j impose constraints on the parameter space, thereby reducing its dimension by q. This is consistent with the theory of Chapter 3 because if it easily verified that M is symmetric and idempotent and therefore, has rank equal to its trace, which is $k - q$. For example, the test statistics have asymptotic χ^2 distributions.

Since it will always be possible to construct a $\hat{B}$ so that $\hat{B}^T \hat{M} \hat{B} = I_p$ for any $p \leq k - q$ (for example, see Anderson 1958, p. 341) we then have, with such a $\hat{B}$,

$$S(\hat{\beta}) = \hat{S}_p = \sum_{i=1}^{p} \hat{V}_i^2 \quad \text{where} \quad \hat{V} = (\hat{V}_i) = \hat{B}^T \sum_{j=1}^{n} h(X_j; \hat{\beta})/\sqrt{n}$$

The asymptotic covariance matrix of $\hat{V}$ is I_p so that the components $\hat{V}_i$ are in this sense orthogonal.

As essentially equivalent derivations apply for discrete distributions, Theorems 6.1.1 and 6.1.2 will henceforth be assumed to hold for discrete as well as continuous distributions.

6.2 Smooth Tests for the Univariate Normal Distribution

The assumption of normality is basic to much statistical theory and practice. Many tests to check the adequacy of this assumption have now been proposed and the first six sections of Mardia (1980) list a large proportion of them. One normality test that seems to have been largely ignored is that proposed by Lancaster (1969, p. 153). We will now derive this test using the results of §6.1.

We apply Theorem 6.1.1 with

$$f(x; \beta) = \exp\{(x - \beta_1)^2/(2\beta_2^2)\}/\{\beta_2\sqrt{(2\pi)}\}, \qquad -\infty < x < \infty$$

If $H_r(z)$ are the normalized Hermite–Chebyshev polynomials, then

$$\int_{-\infty}^{\infty} H_r(z) H_s(z) \exp(-z^2/2)\, dz/\sqrt{(2\pi)} = \delta_{rs}$$

Putting $X = \beta_1 + \beta_2 Z$ in this equation shows that $h_r(x; \beta) = H_r((x - \beta_1)/\beta_2)$, are orthonormal with respect to the distribution of X. Differentiating $\log f(x; \beta)$ gives

$$\partial \log f / \partial \beta_1 = (x - \beta_1)/\beta_2^2 = h_1(x; \beta)/\beta_2$$

and

$$\partial \log f / \partial \beta_2 = (x - \beta_1)^2/\beta_2^3 - 1/\beta_2 = (\sqrt{2}/\beta_2) h_2(x; \beta)$$

since $h_1(x; \beta) = (x - \beta_1)/\beta_2$ and $h_2(x; \beta) = \{(x - \beta_1)^2/\beta_2^2 - 1\}/\sqrt{2}$. Routine calculations show that

$$\mathrm{cov}_0\left(\frac{\partial \log f}{\partial \beta}, \frac{\partial \log f}{\partial \beta}\right) = \begin{pmatrix} \beta_2^{-2} & 0 \\ 0 & 2\beta_2^{-2} \end{pmatrix}$$

and ultimately M is the direct sum of the 2×2 null matrix and the order $k - 2$ unit matrix. So Theorem 6.1.1 fails because M is singular.

One solution is to modify model 6.1.1 by removing the first two orthonormal functions from Equation 6.1.1 and then applying Theorem 6.1.1. Without h_1 and h_2 in the model, $\mathrm{cov}_0(h, \partial \log f/\partial \beta) = 0$ and $M = I_k$. Then the score statistic is

$$S(\hat{\beta}) = \sum_{r=3}^{k} \hat{V}_r^2 = \hat{S}_{k-2} \quad \text{say, with} \quad \hat{V}_r = \sum_{j=1}^{n} h_r(X_j; \hat{\beta})/\sqrt{n}$$

Another alternative is to apply Theorem 6.1.2. A suitable matrix for $\hat{B}$ would have arbitrary first two rows and $(\hat{B})_{rs} = \delta_{r-2,s}$, $r = 3, \ldots, k$, and $s = 1, \ldots, k$. Then $\hat{B}^T\hat{M}\hat{B} = I_k$. Neither h_1 nor h_2 appear in $S(\hat{\beta})$ because the maximum likelihood estimators $\hat{\beta}_1 = \bar{X}$ and $\hat{\beta}_2 = \sum (X_j - \bar{X})^2/n$ are defined by $\sum_{j=1}^n h_r(X_j; \hat{\beta}) = 0$, $r = 1, 2$. Again $\hat{S}_{k-2}$ results.

Critical Values and Power Comparisons

Note that $\hat{S}_1$ will not be considered in the power comparisons as its performance will be like that of the skewness coefficient $\sqrt{b_1}$, which has been studied previously (for example, see Filliben, 1975).

As in Best and Rayner (1985b), approximate powers for tests based on $\hat{S}_2$, $\hat{S}_3$, and $\hat{S}_4$ will be given for

1. a test size (α) of 5% and sample size of $n = 20$, and
2. a test size of 10% and sample size of $n = 90$.

Comparisons can then be made with some of the powers given in Tables 5 and 6 of Stephens (1974a). A Tukey distribution with $\beta = 10$ is shown; a Tukey (β) distribution is given by $X^\beta - (1 - X)^\beta$ where X is uniform on the unit interval. Also included is a scale contaminated alternative denoted by $SC(\beta_1, \beta_2)$. This has a probability density function

$$\{(1 - \beta_1)e^{-x^2/2} + (\beta_1/\beta_2)e^{-x^2\beta_2^2/2}\}/\sqrt{(2\pi)}$$

All the alternatives shown are defined in Table 5 of Pearson et al. (1977). For the convenience of the reader, the powers of the better performed tests from Stephen's tables are reproduced here. These are the tests based on the Shapiro–Wilk statistic W, its modification, the Shapiro–Francia statistic W', the Cramer–von Mises statistic W^2, and the Anderson–Darling statistic A^2.

Even for $n = 90$ use of χ_k^2 critical values for $\hat{S}_k$ are not entirely satisfactory and so values determined from Monte Carlo samples of size 10,000 were used to find approximate critical values for $n = 5$, 10, 15, 20, 25, 35, 50, 70,

90, and 200. For $n \geq 15$, $\alpha = 0.05$ and 0.10, plots against functions of n of the critical values from the Monte Carlo samples suggest that approximate 95% and 90% critical values of $\hat{S}_k$ are obtained by multiplying the corresponding percentage point of χ_k^2 by $(1 - 1.6/\sqrt{n})$ and $(1 - 1.8/\sqrt{n})$, respectively. Thus, for $n = 20$ and $\alpha = 0.10$ the approximate critical value is 2.75 for $\hat{S}_2$ and 4.65 for $\hat{S}_4$. The constants 1.6 and 1.8 were determined by standard regression techniques. Simulated sizes of the tests based on these approximate critical values were generally within 1% of the required 5% or 10% value.

The $\hat{S}_k$ powers given in Table 6.1 are based on Monte Carlo samples of size 200 and use the critical values based on the Monte Carlo samples rather than on the χ^2 approximation. Notice that $\hat{S}_2$ does rather worse than $\hat{S}_4$ for the χ_1^2, exponential and Tukey alternatives where there are deviations from normality other than skewness and Kurtosis. On the other hand, there are no cases where $\hat{S}_2$ does appreciably better than $\hat{S}_4$.

An observation that can be made from Table 6.1b is that the W' test does badly for the Tukey alternative. In fact, unweighted correlation tests do badly for many symmetric alternatives with shorter tails than the normal. See Table 2 of Filliben (1975). Furthermore, Pearson et al. (1977) considered how W' is affected by ties. The components $\hat{Z}_4$ and $\hat{Z}_6$ are particularly effective in detecting alternatives with shorter tails while large $\hat{Z}_3$ and $\hat{Z}_5$ components detect longer tail alternatives.

Table 6.1 Power comparisons for tests of normality

Alternative	$\hat{S}_2$	$\hat{S}_3$	$\hat{S}_4$	$W(W')$	W^2	A^2
(a) $n = 20$, $\alpha = 0.05$						
χ_1^2	0.81	0.95	0.93	0.97	0.94	0.98
χ_4^2	0.40	0.47	0.45	0.46	0.45	0.48
Exponential	0.61	0.74	0.70	0.85	0.74	0.82
χ_{10}^2	0.26	0.28	0.25	0.28	0.23	0.31
Log-normal	0.80	0.88	0.86	0.93	0.88	0.91
Tukey, $\beta = 10$	0.51	0.54	0.65	0.86	0.93	0.90
Cauchy	0.86	0.85	0.89	0.87	0.88	0.98
Laplace	0.31	0.31	0.34	0.25	0.26	0.14
Student-t_4	0.30	0.28	0.30	0.24	0.21	0.23
Student-t_6	0.17	0.16	0.17	0.15	0.12	0.14
SC; $\beta_1 = 0.1$, $\beta_2 = 3$	0.37	0.35	0.36	0.26	0.28	0.30
(b) Power comparison, $n = 90$, $\alpha = 0.10$						
χ_{10}^2	0.88	0.91	0.88	0.85	0.67	0.76
Weibull, $k = 2$	0.70	0.71	0.70	0.75	0.64	0.76
Laplace	0.83	0.81	0.89	0.90	0.86	0.86
Tukey, $\beta = 5$	0.41	0.37	0.77	0.41	0.64	0.62
Student-t_4	0.81	0.76	0.82	0.78	0.66	0.69
SC; $\beta_1 = 0.1$, $\beta_2 = 3$	0.87	0.85	0.86	0.86	0.58	0.68

The results for the scale contaminated normal are interesting. The $\hat{S}_k$ test statistics do rather better than W. Table 6.1b indicates that W^2 and A^2 do not perform as well for this alternative either. We conclude from Table 6.1 that $\hat{S}_4$ is a good omnibus test of normality.

Examples 6.2

Consider the following three data sets of size 20:

1. 53, 1, 70, 73, 79, 48, 91, 20, 34, 91, 87, 15, 3, 78, 78, 62, 6, 15, 20, 42
2. −3, −454, 116, 6, −153, −46, −42, 183, −87, −4, −191, 36, 97, −48, −48, 56, 92, −32, −361, −67
3. 4, 1, 2, 2, 3, 4, 1, 2, 4, 5, 5, 2, 3, 3, 6, 5, 10, 0, 9, 3.

Suppose we choose $\alpha = 0.10$. Calculation gives $\hat{S}_4 = 4.74$, 9.22, 6.71 for (1), (2), and (3) respectively. All of these values exceed the critical values of 4.65. For data set (1) neither $\hat{S}_2$ nor the Filliben correlation tests exceeds the 10% critical value. It appears from consideration of $\hat{S}_4$, then, that there is evidence of nonnormality in each of (1), (2), and (3). The components of $\hat{S}_4$, i.e. $\hat{Z}_3$, $\hat{Z}_4$, $\hat{Z}_5$ and $\hat{Z}_6$, give an indication of the type of nonnormality. The components are:

	$\hat{Z}_3$	$\hat{Z}_4$	$\hat{Z}_5$	$\hat{Z}_6$
(1)	−0.23	−1.36	0.35	1.65
(2)	−2.16	1.19	1.09	−1.39
(3)	1.90	0.70	−1.22	−1.05

For (1), $(\hat{Z}_4^2 + \hat{Z}_6^2)/\hat{S}_4 = 0.96$. Such dominance by the even components is typical of a symmetric alternative with a shorter tail than the normal. For (2), all the components are rather large, which is typical of a symmetric alternative with a longer tail than the normal. For (3), $(\hat{Z}_3^2 + \hat{Z}_5^2)/\hat{S}_4 = 0.76$; dominating odd component or components is typical of a skewed alternative. In this case, a symmetrizing transformation may be helpful, and in fact the square roots of the data of set (3) yield $\hat{S}_4 = 2.1$, a nonsignificant value.

Even if $\hat{S}_4$ is not significant, a check on whether one component or a pair of components are dominant may be suggestive. The data sets (1) and (2) are actually rounded values from random samples from uniform and Laplace distributions. Data set (3) is a random sample from a Poisson distribution. See Figures 6.1 and 6.2 for normal Q–Q plots.

According to the descriptions of common patterns given, for example, in Weisberg (1980), p. 134), the Q–Q plots show;

for data set (1): too many values away from the mean
for data set (2): negative skewness
for data set (3): positive skewness.

For data sets (1) and (3) these descriptions agree with our examination of

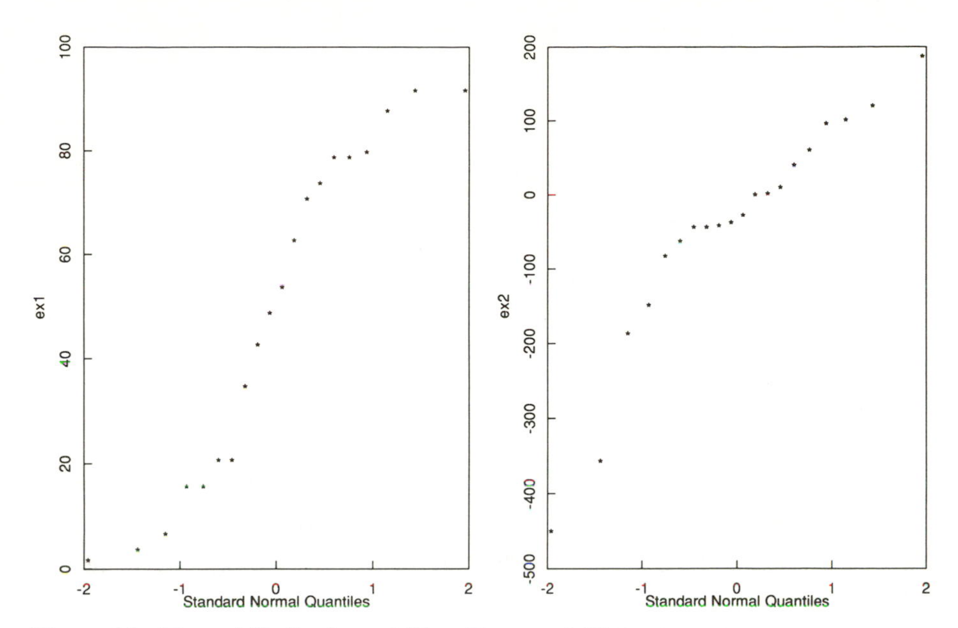

Figure 6.1 Normal Q–Q plots of (1) uniform and (2) Laplace distributed data.

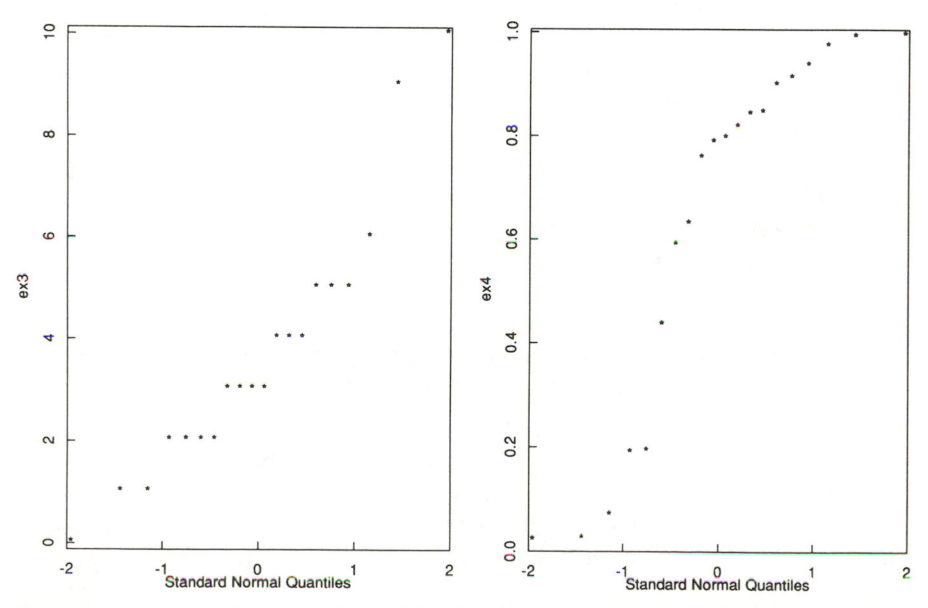

Figure 6.2 Normal Q–Q plots of (3) Poisson distributed data and (4) times of loom breakdowns.

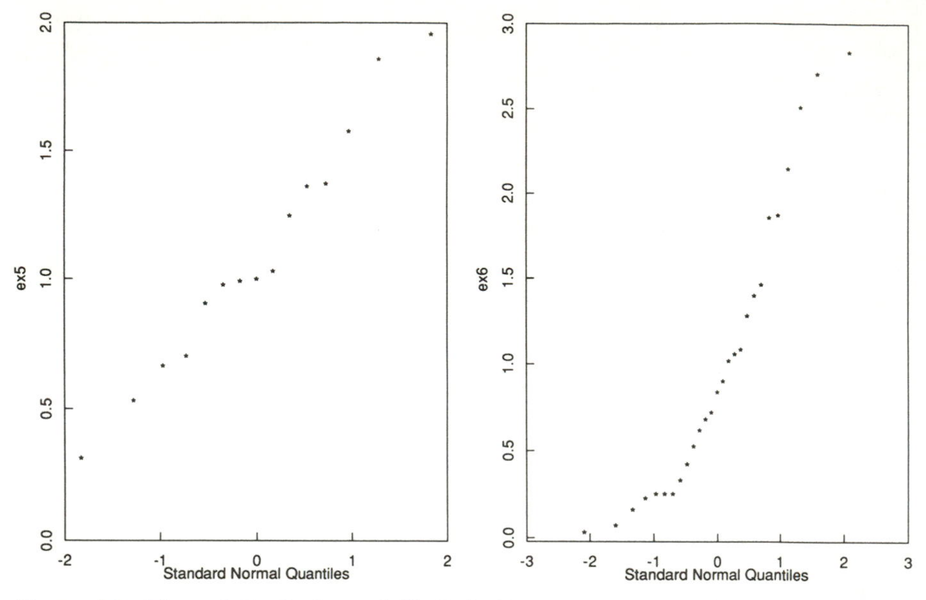

Figure 6.3 Normal Q–Q plots of (5) Rolls-Royce fatigue and (6) air-conditioning failure data.

the components. For data set (2) we observe that heavy-tailed alternatives in small samples often exhibit skewness.

As three further examples of the use of the components let us consider the data given in Pearson and Hartley (1972, pp. 120–121). Only (5), the second of the following sets is actually likely to be normally distributed, but it is instructive to look at all three. The data sets are:

4. 0.0197, 0.0236, 0.0684, 0.1882, 0.1914, 0.4329, 0.5875, 0.6283, 0.7559, 0.7862, 0.7947, 0.8158, 0.8401, 0.8434, 0.8967, 0.9105, 0.9349, 0.9717, 0.9895, 0.9934
5. 0.301, 0.519, 0.653, 0.690, 0.892, 0.964, 0.978, 0.987, 1.017, 1.233, 1.346, 1.357, 1.562, 1.845, 1.944
6. 0.013, 0.052, 0.143, 0.208, 0.234, 0.234, 0.234, 0.312, 0.404, 0.508, 0.599, 0.664, 0.703, 0.820, 0.885, 1.002, 1.041, 1.067, 1.263, 1.380, 1.445, 1.836, 1.849, 2.122, 2.486, 2.682, 2.812.

Data set (4) is derived from times of successive warp-breaks on a loom; data set (5) is derived from rotating bend fatigue data supplied by Rolls-Royce; and data set (6) is derived from days between air conditioning failure in Boeing 720 aircraft. Again, see Figures 6.2 and 6.3 for normal Q–Q plots. Only data set (5) gives a fairly linear plot. The components are:

	$\hat{Z}_3$	$\hat{Z}_4$	$\hat{Z}_5$	$\hat{Z}_6$
(4)	−1.39	−0.89	2.00	0.83
(5)	0.45	−0.50	−0.56	0.20
(6)	1.70	−0.45	−1.87	0.18

For data sets (4) and (6), $(\hat{Z}_3^2+\hat{Z}_5^2)/\hat{S}_4>0.80$ and a skewed alternative is indicated. Use of $\hat{S}_4$ for (4) would not have resulted in the rejection of the null normality hypothesis. Data set (5) is well fitted by the normal, agreeing with the Q–Q plot in Figure 6.3.

Relationship with a Test of Thomas and Pierce

Thomas and Pierce (1979) have also defined tests for this situation, based on quadratic score statistics, W_k. These statistics are based on a nonorthogonal basis and therefore require a table of coefficients to define the test statistics. They are optimal in the same sense as the $\hat{S}_k$, against the same alternatives.

There is likely to be little difference in power between Lancaster's $\hat{S}_k$ statistics and Thomas and Pierce's W_k statistics. In fact, using the powers of W_k given in an unpublished doctoral thesis by Kang (1979), as well as the appropriate figures from our Table 6.1a, we can get the comparison shown in Table 6.2. We claim that the corresponding $\hat{S}_k$ and W_k tests have a Pitman asymptotic relative efficiency 1.0; therefore, there should be no consistently superior test in terms of power. To us the significant differences between these tests are:

1. that the components of the $\hat{S}_k$ are asymptotically independent,
2. that the first one or two components of $\hat{S}_k$ are often well known and powerful tests in their own right.

6.3 Smooth Tests for the Exponential Distribution

Tests of exponentiality are important. For example, see Lin and Mudholker (1980) and the review by Angus (1982). Again we seek to apply the theory of §6.1 with

$$f(x;\beta)=\beta e^{-\beta x},\ x>0,\ \text{zero otherwise, where } \beta>0$$

Suppose $L_r(z)$ are the Laguerre polynomials, orthonormal with respect to $f(z;1)$. These may be defined by

$$L_r(z)=\sum_{s=0}^{r} {}^rC_s(-z)^s/s!$$

Table 6.2 Comparison of Powers for $\hat{S}_k$ and W_k test of normality, $n=20$, $\alpha=0.05$

Alternative	$\hat{S}_2$	W_2	$\hat{S}_3$	W_3	$\hat{S}_4$	W_4
Laplace	0.31	0.33	0.31	0.31	0.34	0.29
Cauchy	0.86	0.90	0.85	0.89	0.89	0.89
Lognormal (0.1)	0.80	0.90	0.88	0.86	0.86	0.88
Exponential ($\Gamma(1)$)	0.61	0.72	0.74	0.65	0.70	0.71

By substituting $X = Z/\beta$ in the orthonormality conditions for $L_r(z)$, we find that $h_r(x;\beta) = L_r(\beta x)$, $r = 0, 1, 2, \ldots$, are orthonormal. Since $h_1(x;\beta) = 1 - \beta x$,

$$\frac{\partial \log f}{\partial \beta} = \frac{1}{\beta} - x = \frac{h_1(x;\beta)}{\beta}$$

Routine calculations show that $\text{cov}_0(\partial \log f/\partial\beta,\ h) = \text{cov}_0(\beta^{-1}h_1, h) = (\beta^{-1}, 0, \ldots, 0)$ and $\text{var}_0(\partial \log f/\partial\beta) = \beta^{-2}$, from which M is the direct sum of the scalar 0, and the $(k-1)\times(k-1)$ unit matrix I_{k-1}. As in the normal case, the solution is to either drop $h_1(x;\beta)$ from model 6.1.1 and apply Theorem 6.1.1, or to apply Theorem 6.1.2 with $\hat{B}$ having arbitrary first row, and subsequent rows defined by $(\hat{B})_{rs} = \delta_{r-1,s}$, for $r = 2, \ldots, k$ and $s = 1, \ldots, k$. In either case, the score statistic is

$$S(\hat{\beta}) = \sum_{r=2}^{k} \hat{V}_r^2 = \hat{S}_{k-1}, \qquad \hat{V}_r = \sum_{j=1}^{n} L_r(\hat{\beta}X_j)/\sqrt{n}$$

The maximum likelihood estimator is $\hat{\beta} = 1/\bar{X}$. The nonappearance of $L_1(.)$ in the score statistic is reasonable, because $\sum_j L_1(X_j/\bar{X}) = 0$. Note that $\hat{V}_2$ is a linear translation of Greenwood's statistic (for example, see Koziol, 1987, Remark 3, p. 22).

Although the Thomas and Pierce statistics W_k and our $\hat{S}_k$ are asymptotically distributed as χ^2_k, it is important for the practical application of the test to assess the rate of approach to this distribution. It has been shown in Rayner and Best (1986) that the expected values of the $\hat{V}_r$, the components of the $\hat{S}_k$, satisfy $E[\hat{V}_2] = -\sqrt{n}/(n+1)$ and, for $r > 2$, $E[\hat{V}_r] = 0(n^{-3/2})$. Convergence should therefore be reasonably rapid, although not as rapid as in the uniform case (which was considered by Solomon and Stephens, 1983). To verify this, and to find usable finite sample critical points, a Monte Carlo experiment was performed. Samples of size 10,000 were sorted into ascending order for $n = 5$, 10, 15, 20, 30, 50, 100, and 200. Reasonable approximate critical values for $n \geq 10$ were obtained for $\hat{S}_k (k = 2, 3, 4)$ for $\alpha = 0.05$ and 0.10 by multiplying the χ^2_k critical value by $(1 - 1.5/\sqrt{n})$ and $(1 - 1.8/\sqrt{n})$, respectively. Our investigation suggests that when using these critical values the size of the test is in error by no more than 1%.

From previous power studies it would appear that, while no omnibus test is always most powerful, the Gini statistic, G_r (Gail and Gastwirth, 1978a) performs very well. Given ordered observations $X_{(1)} < \ldots < X_{(n)}$,

$$G_r = \left[\sum_{i=1}^{n-1} \{i(r-i)(X_{(i+1)} - X_{(i)})\}\right] \Big/ \left\{(r-1)\sum_{i=1}^{r} X_i\right\}$$

Table 6.3 compares the power of G_{20} with that of $\hat{S}_2 = \hat{V}_2^2 + \hat{V}_3^2$ and $\hat{S}_4 = \hat{V}_2^2 + \hat{V}_3^2 + \hat{V}_4^2 + \hat{V}_5^2$ for $\alpha = 0.05$ and 0.10. Results for $\hat{S}_1$ and $\hat{S}_3$ were both inferior to those for $\hat{S}_2$ and $\hat{S}_4$ and are not shown. The probability density functions for the alternatives used are described by Gail and

Table 6.3 Power comparisons of tests for exponentiality; $n = 20$

Alternative	G_{20}	$\hat{S}_2^2$	$\hat{S}_4^2$	W_2
(a) $\alpha = 0.1$				
χ_1^2	0.662	0.601	0.668	0.667
χ_3^2	0.287	0.257	0.228	0.256
χ_4^2	0.602	0.557	0.496	0.549
χ_8^2	0.992	0.983	0.983	0.992
Log normal (0.6)	0.887	0.790	0.877	0.127
Log normal (0.8)	0.338	0.301	0.381	0.159
Log normal (1.0)	0.185	0.234	0.272	0.177
Log normal (1.2)	0.372	0.422	0.436	0.189
Weibull (0.5)	0.940	0.917	0.928	0.966
Weibull (2.0)	0.981	0.977	0.955	0.960
Beta (1, 2)	0.406	0.402	0.272	0.260
$(\chi_{0.5}^2 + \chi_4^2)/2$*	0.604	0.503	0.748	0.922
$(\chi_1^2 + \chi_5^2)/2$*	0.187	0.142	0.291	0.531
(b) $\alpha = 0.05$				
Weibull (0.8)	0.239	0.243	0.234	0.271
Weibull (1.5)	0.505	0.382	0.323	0.329
Uniform (0, 2)	0.722	0.671	0.502	0.462
Pareto (3)	0.798	0.687	0.925	0.997
Shifted Pareto (3)	0.470	0.490	0.476	0.418
Shifted exponential (0.2)	0.230	0.172	0.170	0.276
$(\chi_{0.5}^2 + \chi_4^2)/2$*	0.453	0.402	0.649	0.888
$(\chi_1^2 + \chi_5^2)/2$*	0.110	0.089	0.210	0.438

Gastwirth (1978b, Table 3) and by Angus (1982). If $F(x)$ is the appropriate cumulative density function, then 1,000 Monte Carlo samples, each with $n = 20$, were generated from the shifted exponential, Pareto and shifted Pareto distributions by solving $u = F(x)$ for x, where u is a random value from a uniform (0, 1) distribution. For the uniform, Weibull, χ^2, beta and log normal distributions, the samples of size 1,000 were generated using routines from the IMSL (1982) library. The values of n and α were chosen so that most powers in Table 6.3 may be compared with those shown in Tables 2 and 4 of Angus (1982). Thus, comparisons with a wide selection of previously suggested test statistics can be made. Some of the alternatives, marked in Table 6.3 with an asterisk, were chosen because the $\hat{S}_k$ and W_k tests were likely to do well. As our interest is in omnibus tests, all tests are two-sided. It appears that $\hat{S}_2$, $\hat{S}_4$, and W_2 compare favorably as omnibus tests for exponentiality.

An advantage of the $\hat{S}_k$ statistics is that their components may indicate what alternative distribution would fit the data. Using the components in this data analytic manner requires some experience on the part of the user.

Consider now some data of Angus (1982) who gave 20 operational

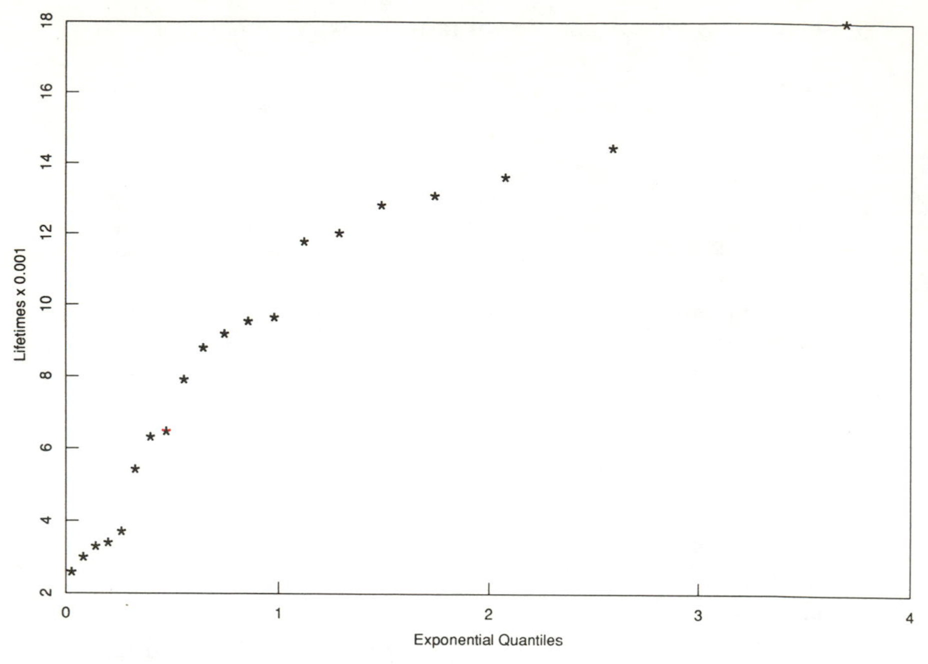

Figure 6.4 Exponential Q–Q plots of lifetimes.

lifetimes in hours:

> 6,278, 3,113, 5,236, 11,584, 12,628, 7,725, 8,604, 14,266, 6,125, 9,350, 3,212, 9,003, 3,523, 12,888, 9,460, 13,431, 17,809, 2,812, 11,825, 2,398.

Suppose the data are tested for exponentiality with the mean unspecified. An exponential Q–Q plot is given in Figure 6.4, and casts doubt on the possibility of exponential lifetimes. We calculate $\hat{S}_4 = 8.8$, which is larger than the appropriate 95% critical point 6.3. This is in agreement with the

Table 6.4 Values of $E[\hat{V}_r]$ for χ^2 alternatives when $n = 20$

Alternative	$E[\hat{V}_2]$	$E[\hat{V}_3]$	$E[\hat{V}_4]$	$E[\hat{V}_5]$
χ_1^2	1.63	0.88	0.70	0.75
Null (χ_2^2)	−0.21	−0.04	0.02	0.01
χ_3^2	−0.87	−0.75	−0.50	−0.32
χ_4^2	−1.20	−1.21	−0.89	−0.56
χ_5^2	−1.40	−1.51	−1.17	−0.75
χ_6^2	−1.54	−1.72	−1.38	−0.90
χ_7^2	−1.64	−1.88	−1.54	−1.03
χ_8^2	−1.71	−2.01	−1.67	−1.13
χ_9^2	−1.77	−2.11	−1.78	−1.21
χ_{10}^2	−1.82	−2.19	−1.87	−1.28

conclusion reached by Angus, but we also have $(\hat{V}_2, \hat{V}_3, \hat{V}_4, \hat{V}_5) = (-1.7, -1.9, -1.4, -0.8)$. Now

$$E[\hat{V}_r]/\sqrt{n} = \sum_{s=0}^{r} {}^rC_s(-1)^s E[(X_1/\bar{X})^s)]/s!$$

If a χ_s^2 alternative is indicated then, it may be shown that

$$E[(X_1/\bar{X})^s] = n^s\{\Gamma(s+k/2)/\{\Gamma(k/2)\}\{\Gamma(s+nk/2)/\Gamma(nk/2)\}^{-1}$$

For a range of values of k and for $n = 20$, values of $E[\hat{V}_r]$ are given in Table 6.4. This table suggests a χ_7^2 distribution as an alternative life model. For alternatives where it is difficult to find $E[(X_1/\bar{X})^s]$, Monte Carlo methods can be used.

6.4 Smooth Tests for the Poisson Distribution

The Poisson distribution is one of the better-known discrete probability distributions and has many applications (for example, see the monograph by Haight, 1967, or Johnson and Kotz, 1969, Chapter 4). The Poisson probability function is

$$f(x; \lambda) = e^{-\lambda}\lambda^x/x!, \qquad x = 0, 1, 2, \ldots, \text{ where } \lambda > 0$$

Two well-known goodness of fit tests for the Poisson distribution are Karl Pearson's classic chi-squared test, and a test due to R.A. Fisher based on the index of dispersion,

$$D = \sum_{j=1}^{n} (X_j - \bar{X})^2/\bar{X}$$

in which $\bar{X}$ is the mean of a random sample of $X_1, \ldots, X_n$ from the distribution hypothesized to be Poisson. The validity of the χ^2 approximation to the distribution of the Pearson test statistic, however, is uncertain, as is the amount of pooling required. We will not, therefore, consider Pearson's test further, although we note that Svensson (1985) discussed alternative tests for pooled data.

We will now develop a smooth test statistic for the Poisson distribution based on the Poisson–Charlier orthonormal polynomials. The components of this statistic are asymptotically independent and may be interpreted as identifying deviations of the data from the Poisson moments. Gart (1975) also proposed statistics to identify such deviations, but did not explore small sample properties.

Consider the model

$$C(\theta, \lambda) \exp\left\{\sum_{i=1}^{k} \theta_i h_i(x; \lambda)\right\} e^{-\lambda}\lambda^x/x!$$

where $h_i(x; \lambda) = \sqrt{(\lambda^i/i!)}\sum_{v=0}^{i} (-1)^{i-v}\, {}^iC_v\, v!\lambda^{-v}\, {}^xC_v$ are the Poisson–Charlier orthonormal polynomials, discussed, for example, in Szego (1959, § 2.81). Of course $\theta = (\theta_1, \ldots, \theta_k)^T$ and $C(\theta, \lambda)$ is a normalizing constant.

Routine calculations show that the score vector has asymptotic covariance matrix n times the matrix $\text{diag}(0, 1, 1, \ldots, 1)$. This uses $h_1(x; \lambda) = (x - \lambda)/\sqrt{\lambda}$ so that when $\theta = 0$, $\text{cov}(h_r, X) = \text{cov}(h_r, \sqrt{\lambda} h_1 + \lambda) = 0$ for $r > 1$. Possible solutions are to either remove $h_1(x; \lambda)$ from the model and use Theorem 6.1.1, or to use Theorem 6.1.2. The former is reasonable because the singularity of the asymptotic covariance matrix shows that asymptotically, the first component of the score vector is dependent upon the other components.

The usable score statistic for testing $H: \theta = 0$ against $K: \theta \neq 0$ with $h_1(x; \lambda)$ removed, is, for this model, based on

$$\hat{S}_{k-1} = \sum_{i=2}^{k} \hat{V}_i^2, \quad \text{in which} \quad \hat{V}_i = \sum_{j=1}^{n} h_i(X_j; \hat{\lambda})/\sqrt{n} \quad \text{and} \quad \hat{\lambda} = \bar{X}$$

Notice that

1. $\hat{V}_2 = (D - n)/\sqrt{(2n)}$ is just a standardized version of D
2. $E[\hat{V}_2] = -(2n)^{-1/2}$ if $D = (n - 1)$ when $\hat{\lambda} = 0$
3. In the Charlier type B approximation, $g(x; \hat{\lambda})$ is given by

$$g(x; \hat{\lambda}) = f(x; \hat{\lambda}) \left\{ 1 + \sum_{i=2}^{k} \hat{V}_i h_i(x; \hat{\lambda})/\sqrt{n} \right\}$$

Definition (2) is made to avoid having $\hat{V}_2$ indeterminate when $\hat{\lambda} = 0$. The Charlier type B series could be viewed as resulting from approximating e^x by $1 + x$, which is reasonable if $x = O(n^{-0.5})$. Barton (1953) used this device. Kendall and Stuart (1977, §6.24) discussed Type B series. The problem of the choice of terms in this series is one reason why type B series density estimators are not much used. As a first step toward a solution to this problem, observe that the $\hat{V}_i$ are asymptotically independent standard normal. An immediate guide in the choice of terms in $g(x; \hat{\lambda})$ would be to ignore those $\hat{V}_i$ that are significantly small, say $|\hat{V}_i| < 2$. In a goodness of fit context, any negative frequencies derived from $g(x; \hat{\lambda})$ are an indication that the actual underlying distribution is not Poisson-like. Examination of the $\hat{V}_i$ may give an indication of more appropriate probability models.

Subsequently, k is taken to be 5, so that asymptotically $\hat{S}_4$ will have a χ_4^2 distribution. The small sample distribution of $\hat{S}_4$, however, depends on both n and λ. To investigate this behavior, a Monte Carlo experiment using 10,000 samples of size n was carried out for various values of λ and nominal test size α. Table 6.5 gives some results comparing test sizes using χ_4^2 critical values with sizes using corrected χ_4^2 critical values. For $\alpha = 0.05$ the correction consisted of taking the critical value of χ_4^2 multiplied by $\{1 - 1.14/\sqrt{(n\bar{x})}\}$. The correction is useful when $n\bar{x} > 10$ and was found using standard regression techniques to relate the percentage points of $\hat{S}_4$, derived from Monte Carlo experiments, against λ. In the Monte Carlo

Table 6.5 Approximate sizes of $\hat{S}_4$ based on corrected and uncorrected (in brackets) χ_4^2 critical values*

n/λ	0.5	1	3	5	10	20
10	0.08 (0.01)	0.06 (0.02)	0.05 (0.03)	0.06 (0.04)	0.05 (0.04)	0.04 (0.04)
20	0.06 (0.02)	0.05 (0.03)	0.05 (0.03)	0.05 (0.04)	0.05 (0.04)	0.05 (0.04)
30	0.05 (0.03)	0.05 (0.03)	0.05 (0.04)	0.05 (0.04)	0.05 (0.05)	0.05 (0.05)
40	0.06 (0.03)	0.05 (0.03)	0.05 (0.04)	0.05 (0.04)	0.06 (0.04)	0.06 (0.05
100	0.05 (0.04)	0.05 (0.04)	0.05 (0.04)	0.05 (0.05)	0.06 (0.05)	0.05 (0.05)

* Various λ and n are shown and the nominal size is $\alpha = 0.05$.

investigations, use was made of the recurrence relation for the Poisson–Charlier polynomials given in Appendix 2.

Table 6.6 gives a brief power comparison for tests based on $\hat{S}_4$ with those based on the index of dispersion D and a test of Campbell and Oprian (1979) based on the Kolmogorov–Smirnov statistic, K. As $\hat{S}_4$ is a score statistic, it will have asymptotically optimal power properties. Furthermore, D is also a score statistic for particular alternatives; see Selby (1965), and Rayner and McIntyre (1985). For D, we used a two-tailed test as in Gbur (1981) and χ_{n-1}^2 critical points, which Bennett and Birch (1974) showed were adequate. For K, we used critical values given by Campbell and Oprian (1979) and the corrected χ_4^2 values discussed previously for $\hat{S}_4$.

Routines from the IMSL (1982) library were used to generate 1,000 samples each of 20 points for the discrete alternatives shown in Table 6.6. Clearly, $\hat{S}_4$ and D perform better than K. As D is essentially the first component of $\hat{S}_4$ it would be expected that $\hat{S}_4$ would perform better than D

Table 6.6 Approximate powers of $\hat{S}_4$, D, and K for the alternatives shown and with $n = 20$, $\alpha = 0.05$

Alternative	$\hat{S}_4$	D	K
Negative binomial ($p = 2/3$, $k = 2$)	0.30	0.28	0.16
geometric ($p = 1/2$)	0.58	0.17	0.55
discrete uniform ($k = 3$)	0.75	0.97	0.65
discrete uniform ($k = 10$)	0.48	0.29	0.42
binomial ($n = 3$, $p = 1/2$)	0.15	0.39	0.29
0.5 Poisson ($\lambda = 2$) + 0.5 Posisson ($\lambda = 6$)	0.68	0.68	0.53
Poisson ($\lambda = 3$)	0.05	0.05	0.06

whenever deviations other than in variance were important. If only variance deviations were important, then D would be more powerful than $\hat{S}_4$ because the effect of the first component of $\hat{S}_4$ would be diluted.

Example 6.4.1

Hoaglin (1980) presented the frequency count data shown in Table 6.7. The data were originally given by Rutherford and Geiger and relate to the radioactive decay of polonium. Hoaglin (1980, Figures A and D) presented two plots, which suggest deviations from the Poisson model. The values of $\hat{S}_4$, D, and K for this data are 10.6, 2488.9, and 0.011 with corresponding approximate P values of .03, .10, and .20. The components of $\hat{S}_4$ take the values -1.65, 0.16, 2.81, and -0.09, indicating that the observations have second moment less, and fourth moment greater, than might be expected. In this example, $\hat{S}_4$ has a lower associated P value than D as there are departures from the Poisson model other than in variance.

As $\hat{V}_5$ is an insignificant -0.09, k was chosen as 4 for the sum in $g(x; \hat{\lambda})$. Aroian (1937) also gave these $ng(x; \hat{\lambda})$ values. The $ng(x; \hat{\lambda})$ frequencies are a significant improvement over the $f(x; \hat{\lambda})$ frequencies.

Figure 6.5(a) emphasizes that as we approach a count of 3 the model slightly overestimates the count, then underestimates, then overestimates again. Furthermore, the Poisson predicted value at a count of 8 is somewhat more than the observed.

Chapter 9 of Hoaglin et al. (1985) gave a number of other graphic

Table 6.7 Radioactive decay counts of Polonium with rounded expected frequencies

Count	Frequency	$nf(x; \hat{\lambda})$	$ng(x; \hat{\lambda})$
0	57	54	58
1	203	210	200
2	383	407	386
3	525	525	524
4	532	508	531
5	408	394	418
6	273	254	262
7	139	141	134
8	45	68	58
9	27	29	23
10	10	11	9
11	4	4	4
12	0	1	2
13	1	1	1
14	1	0	0

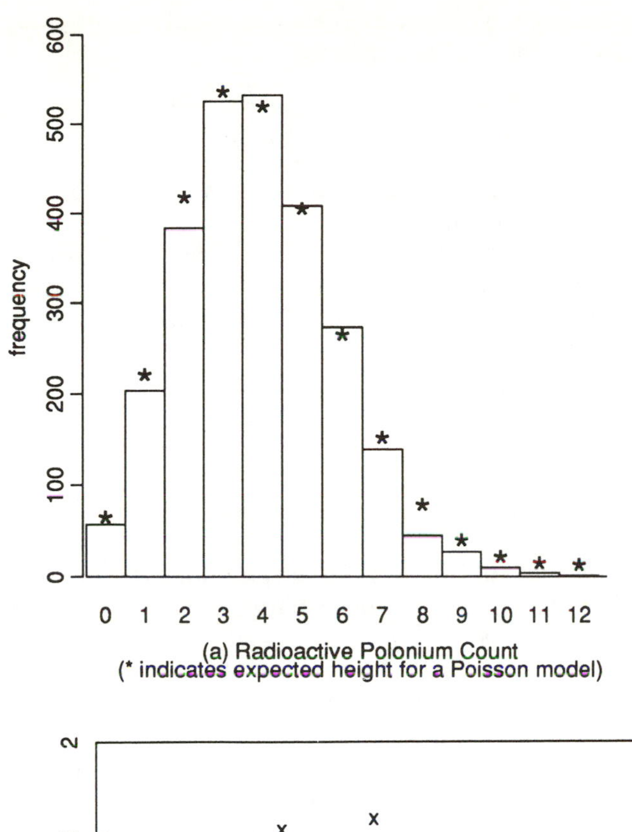

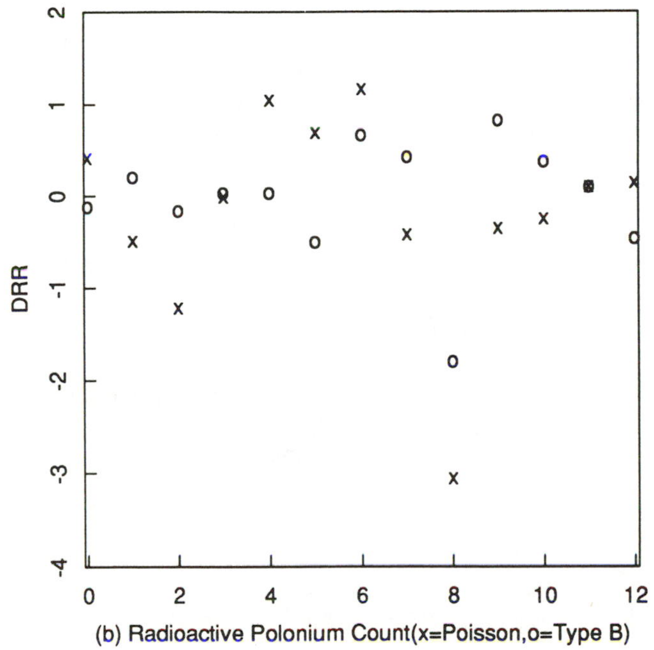

Figure 6.5 Geiger counter data.

displays of this data set. Figure 6.5(b) illustrates one of these—a *double-root residual* (DRR) plot. Define DRR by

$$\begin{aligned} \text{DRR} &= \sqrt{\{4\,\text{Obs} + 2\}} - \sqrt{\{4\,\text{Exp} + 1\}} \text{ for Obs} \geq 1 \\ &= 1 - \sqrt{\{4\,\text{Exp} + 1\}} \text{ for Obs} = 0 \end{aligned}$$

where Obs = the observed count and Exp = the expected count. The DRR are approximately independent $N(0, 1)$ variables.

The points observed from Figure 6.5(a) concerning the trends and the difference at the count of 8 are evident in Figure 6.5(b) also. The type B fit given by column $ng(x; \hat{\lambda})$ in Table 6.7 has DRR values less scattered about the line $y = 0$ in Figure 6.5(b) than those given by the column $nf(x; \hat{\lambda})$. Also, from the DRR values for the column $ng(x; \hat{\lambda})$, the frequency corresponding to the 8 count is still rather small.

6.5 Smooth Tests for the Geometric Distribution

Let X be the number of failures before a success in a sequence of independent trials with common probability, p, of success. Then the probability function of X is $f(x; q) = pq^x$, $x = 0, 1, 2, \ldots ; q = (1 - p)$, and X is said to have a geometric distribution. The geometric distribution describes a number of phenomena and is the discrete analogue of the exponential distribution.

Goodness of fit for the geometric distribution is usually tested using Pearson's X^2 test. However, the validity of the χ^2 approximation to the distribution of the Pearson test statistic is uncertain, as is the amount of pooling required. Another possibility is to use the dispersion type test, $D = X_1^2 + \ldots + X_n^2$, suggest by Vit (1974). A graphic check was given by Hoaglin et al. (1985).

We assume as usual that a random sample $X_1, \ldots, X_n$ is available from the distribution hypothesized to be geometric. Consider the model

$$g_k(x; \theta, \beta) = C(\theta, \beta) \exp\left\{\sum_{r=1}^{k} \theta_r h_r(x; \beta)\right\} f(x; \beta)$$

where $f(x; q) = pq^x$. For the geometric case β consists of q alone. Its maximum likelihood estimator is $\hat{q} = \bar{X}/(1 + \bar{X})$ and $h_r(x; q)$ are a special case of the Meixner (1934) polynomials. The first five are:

$$h_1(x) = (x - a)$$

$$h_2(x) = x(x - 1) - 4ax + 2a^2$$

$$h_3(x) = x(x - 1)(x - 2) - 9ax(x - 1) + 18xa^2 - 6a^3$$

$$h_4(x) = 24\,{}^xC_4 - 96a\,{}^xC_3 + 144a^2\,{}^xC_2 - 96xa^3 + 24a^4$$

$$h_5(x) = 120\,{}^xC_5 - 600a\,{}^xC_4 + 1200a^2\,{}^xC_3 - 1200a^3\,{}^xC_2 + 600a^4x - 120a^5$$

where $a = q/p$. To normalize, divide each $h_r(x)$ by $\sqrt{\{(r!)^2(a^2+a)^r\}}$. A recurrence relation for the polynomials in the Meixner class, which includes the geometric, is given in Appendix 1. The geometric formula is given under Part 5.

As with other models, the smooth test can be derived using either Theorem 6.1 or Theorem 6.2. To use the former, because $\sum_j h_1(x_j; \hat{q}) = 0$, $h_1(x; \beta)$ should be omitted from Equation 6.1.1. If this is done, it is easy to show that M is the $(k-1) \times (k-1)$ unit matrix, and the appropriate score statistic is

$$\hat{S}_k = \sum_{r=2}^{k+1} \hat{V}_r^2$$

in which the rth component is $\hat{V}_r = \sum_j h_r(X_j; \hat{q})/\sqrt{n}$.

Analogous to a Charlier Type B series we can express an empirical distribution $g(x; \hat{q})$, as

$$g(x; \hat{q}) = f(x; \hat{q})\left\{1 + \sum_{r=2}^{k+1} \hat{V}_r h_r(x; \hat{q})/\sqrt{n}\right\}$$

If the orthonormal polynomials can be easily calculated by recurrence, then it is not difficult to calculate $\hat{C} = \sum_{r=2}^{N+1} \hat{V}_r^2$, for some suitably large N, say 20. As the $\hat{V}_r$ are asymptotically distributed as independent standard $N(0, 1)$ variables, we have the asymptotic partition of $\hat{C}$ into $\hat{S}_k$ and $(\hat{C} - \hat{S}_k)$. Asymptotically, $\hat{S}_k$ has a χ^2_k distribution, $(\hat{C} - \hat{S}_k)$ has a χ^2_{N-k} distribution, and $\hat{C}$ has a χ^2_N distribution. Alternatively, we can test, as usual, using $\hat{S}_k$ with some predetermined k. If further information is available about the values of further components, this information is supplementary to the formal significance test. It may only be used informally.

The first component of $\hat{S}_k$, $\hat{V}_2$, should perform similarly to Vit's (1974) D statistic. Power comparisons of $\hat{S}_k$ and D, therefore, should show similar trends to those for the Poisson case considered in the previous section. D would therefore be better for alternatives involving only variance differences, and $\hat{S}_k$ would be better otherwise.

Subsequently, k is taken to be 4, so that asymptotically $\hat{S}_4$ will have a χ^2_4 distribution. The small sample distribution of $\hat{S}_4$, however, depends both on n and q. To investigate this behavior, a Monte Carlo experiment using 10,000 samples of size n was carried out for various values of q and nominal test size α. Table 6.8 gives some results comparing actual test sizes using χ^2_4 critical values with sizes using a corrected χ^2 critical value. For $\alpha = 0.05$, the correction consisted of taking the critical value of χ^2_4, 9.488 multiplied by $\{1 + 3.643/n - 2.314/\sqrt{n} - 0.447/\sqrt{(n\hat{q})}\}$, where nq was taken as n when $n\hat{q} < 1$. The correction was found using standard regression techniques to relate the percentage points of $\hat{S}_4$, derived from Monte Carlo experiments, and q. Approximate P values for $\hat{S}_4$ can be found by simulation.

Table 6.8 Approximate sizes of $\hat{S}_4$ based on 1,000 Monte Carlo trials for corrected and, in brackets, uncorrected χ_4^2 values, for various n and q, and $\alpha = 0.05$

n/q	.1	.3	.4	.6	.7	.9
10	.04 (.00)	.06 (.00)	.06 (.01)	.04 (.01)	.04 (.01)	.04 (.01)
20	.03 (.00)	.04 (.02)	.05 (.01)	.05 (.01)	.04 (.01)	.05 (.02)
50	.04 (.02)	.05 (.03)	.06 (.02)	.04 (.02)	.04 (.02)	.05 (.03)
100	.05 (.03)	.05 (.03)	.05 (.04)	.05 (.04)	.05 (.05)	.06 (.03)

Example 6.5.1

Hoaglin et al. (1985) in their Table 9.26 give the following library circulation data:

Number of borrowings	1	2	3	4	5	6	7
Number of books	65	26	12	10	5	3	1

If $X + 1 =$ number of borrowings, then $\hat{q} = \bar{x}/(1 + \bar{x}) = 0.4979$, $\hat{V}_2 = -0.14$, $\hat{V}_3 = -0.98$, $\hat{V}_4 = 1.25$, $\hat{V}_5 = -0.84$, and $\hat{S}_4 = 3.25$. This is nowhere near the approximate 5% critical value of 7.2, and as none of the $|\hat{V}_i|$ are greater than 2 or individually account for most of $\hat{S}_4$, we conclude that the geometric fit is acceptable.

Example 6.5.2

Hoaglin et al. (1985), in their Table 9.13, also give the library circulation data presented here in Table 6.9.

If we again take $X + 1 =$ number of borrowings, we find $\hat{q} = \bar{x}/(1 + \bar{x}) = 0.474597$ and $\hat{V}_2 = 30.2$, $\hat{V}_3 = -14.4$, $\hat{V}_4 = 4.6$, $\hat{V}_5 = -1.8$. Clearly the geometric distribution is not a good fit. The huge positive $\hat{V}_2$ value indicates the observed distribution has a higher variance than expected for a geometric distribution. Hoaglin et al. (1985) also conclude, on the basis of their Figure 9.12(a), that a geometric distribution is not appropriate.

6.6 Smooth Tests for the Multivariate Normal distribution

Excellent reviews of goodness of fit tests for multivariate normality were given by Gnanadesikan (1977), Cox and Small (1978), and Mardia (1980).

Table 6.9 Library circulation data

Number of borrowings	Number of books
1	63,526
2	25,653
3	11,855
4	6,055
5	3,264
6	1,727
7	931
8	497
9	275
10	124
11	68
12	28
13	13
14	6
15	9
16	4

Csorgo (1986) updated these reviews and derived a "formal conservative large sample Kolmogorov-type test" (Csorgo, 1986, p. 709). Other recent works include that of Machado (1983) and Mardia and Foster (1983), who investigated multivariate skewness and kurtosis tests, and Srivastava and Hui (1987) who discussed a generalized Shapiro–Wilk test.

Koziol (1986) gave an excellent compendium of assessments of multivariate normality; however, he gave smooth tests based on statistics $\hat{U}_3^2$ and $\hat{U}_4^2$, saying

> these smooth tests are not quadratic score statistics but instead are immediate analogues of the corresponding smooth tests in the completely specified null case. (Koziol, 1986, p. 2772).

We will show that Koziol's smooth tests are based on quadratic score statistics. The model is formed by imbedding the multivariate normal probability density function in a Neyman-type alternative. Perhaps Koziol was misled by Cox and Small (1978, p. 263):

> the absence of a simple yet general family of distributions extending the multivariate normal precludes the use of a likelihood ratio test.

In fact, Bera and John (1983) also derived tests for multivariate normality using the quadratic score statistic and a Pearson family of alternatives. They claimed that

> since our statistics are derived from the score principle, they are locally most efficient for Pearson alternatives in large samples (Bera and John, 1983, p. 109).

We can make the same claim for our statistics within the smooth family of alternatives.

The multivariate normal distribution of full rank has probability density function

$$f(x;\beta) = (2\pi)^{-p/2}\,|\Sigma|^{-1/2}\exp\{-(x-\mu)^T\Sigma^{-1}(x-\mu)/2\}$$

where $x \in R^p$, p dimensional Euclidean space, and where β is a vector of parameters containing the elements of $\mu = (\mu_r)$ and $\Sigma^{-1} = \Psi = (\psi_{rs})$; β is defined by

$$\mu_r = (\beta)_r, \text{ with } r = 1, \ldots, p \text{ and } \psi_{rs} = (\beta)_{rp+s}, \text{ with } r \text{ and } s = 1, \ldots, p$$

We write X is distributed as $N_p(\mu, \Sigma)$. In order to define an appropriate family of alternatives, first put $Y = A(x-\mu)$ where $A\Sigma A^T = I_p$. Then Y is distributed as $N_p(0, I_p)$, and the elements $Y_1, \ldots, Y_p$ of Y are independent standard normal variates. Note for a given Σ, more than one such A may exist. This transformation is often used in constructing goodness of fit tests for the multivariate normal, especially with $A = \Sigma^{-1/2}$ (for example, see Gnanadesikan, 1977 and Bera and John, 1983). We note that if B is a positive semidefinite matrix, then B^p is the matrix whose eigenvectors are the same as those of B, but whose eigenvalues are those of B raised to the power p.

Suppose now that $\{H_r(z)\}$ are the normalized Hermite–Chebyshev polynomials, orthonormal on the standard normal distribution. So, if $\varphi(z) = \exp(-z^2/2)/\sqrt{(2\pi)}$, then

$$\int_{-\infty}^{\infty} H_r(z)H_s(z)\varphi(z)\,dz = \delta_{rs}$$

Now if we define

$$L_r^*(y) = \prod_{i=1}^{p} H_r(y_i) \tag{6.6.1}$$

then

$$E[L_r^*(Y)L_s^*(Y)] = \prod_{i=1}^{p} E[H_r(Y_i)H_s(Y_i)] = \prod_{i=1}^{p} \delta_{rs} = \delta_{rs}$$

Alternatively, we may define

$$L_{r_1,\ldots,r_p}(y) = H_{r_1}(y_1)\ldots H_{r_p}(y_p) \tag{6.6.2}$$

These L's would normally be ordered by the degree $r = r_1 + \ldots + r_p$, so that the degree r L's are considered before the degree $r+1$ L's. Suppose some ordering has been imposed. Call this ordered system $\{L_r(y)\}$. Both $\{L_r(y)\}$

and $\{L_r^*(y)\}$ are orthonormal on the distribution of Y. It follows that both $\{L_r(A(x-\mu))\}$ and $\{L_r^*(A(x-\mu))\}$ are orthonormal on the distribution of X. Note that A and the orthonormal system, and hence the tests we derive, are not unique. Also $L_0(y) = L_0^*(y) = 1$, implying that every other member of both orthonormal systems has zero expectation.

Henceforth, we will direct our attention to $\{L_r(y)\}$ as defined by Equation 6.6.2. This is the system discussed by Koziol (1986) and for which power comparisons and examples are given by Best and Rayner (1988). Note that there are p orthonormal functions of degree 1, $p + p(p-1)/2 = {}^{p+1}C_2$ orthonormal functions of degree 2, and ${}^{p+r-1}C_r$ orthonormal functions of degree r. If all functions of degree up to r are included in the orthonormal set, then k, the order of the alternative in the following definition is

$$k = {}^{p}C_1 + {}^{p+1}C_2 + \ldots + {}^{p+r-1}C_r = {}^{p+r}C_r - 1$$

which increases rapidly with r.

We will now define an alternative of order k to $f(x;\beta)$ by

$$g_k(x;\theta,\beta) = C(\theta,\beta)\exp\left\{\sum_{r=1}^{k}\theta_r L_r(\Sigma^{-1/2}(x-\mu))\right\} f(x;\beta)$$

The (quadratic) score statistic for such alternatives can be found using Theorem 6.1.1 or Theorem 6.1.2. We will now derive the quantities needed to apply this theorem. It is not difficult to show that

$$2\log f = -p\log(2\pi) + \log|\Psi| + (x-\mu)^T\Psi(x-\mu),$$

so that

$$\partial\log f/\partial\mu_r = \sum_j x_j\psi_{jr} - \mu_r\psi_{rr},\ r = 1,\ldots,p$$

$$2\,\partial\log f/\partial\psi_{rr} = \{\Sigma - (X-\mu)(X-\mu)^T)\}_{rr},\ r = 1,\ldots,p$$

and

$$\partial\log f/\partial\psi_{rs} = \{\Sigma - (X-\mu)(X-\mu)^T)\}_{rs},\ r \text{ and } s = 1,\ldots,p \text{ but } r \neq s.$$

Now write $L_s^{(r)}(y)$ for any $L_s(y)$ of degree r. It follows that

$$\begin{aligned} E[\{\partial\log f/\partial\mu_r\}L_s^{(t)}(Y)] &= E[(\Psi X)_r L_s^{(t)}(\Sigma^{-1/2}(X-\mu))] \\ &= E[(\Sigma^{1/2}Y)_r L_s^{(t)}(Y)] \end{aligned}$$

since

$$E[L_s^{(r)}(Y)] = 0 \quad \text{for} \quad r = 1, 2, \ldots, \sum_j X_j\psi_{jr} = (\Psi X)_r \quad \text{and} \quad X - \mu = \Sigma^{1/2}Y$$

Thus

$$E[\{\partial\log f/\partial\mu_r\}L_s^{(t)}(Y)] = 0, \quad \text{for} \quad t \neq 1$$

using the orthogonality, for no $L_s^{(t)}(y)$, $t \neq 1$, includes a term Y_r

$(r = 1, \ldots, p)$ alone. Also

$$\begin{aligned}-2E[\{\partial \log f/\partial\psi_{rr}\}L_s^{(t)}(Y)] &= E[\{(X_r - \mu_r)^2\}\{L_s^{(t)}(\Sigma^{-1/2}(X-\mu))\}] \\ &= E[(\Sigma^{1/2}YY^T\Sigma^{1/2})_{rr}L_s^{(t)}(Y)] \\ &= E[(a_1Y_1 + \ldots + a_pY_p)^2L_s^{(t)}(Y)](\text{say}) \\ &= 0 \text{ for } t \neq 2,\end{aligned}$$

and similarly

$$E[\{\partial \log f/\partial\psi_{rr}\}L_s^{(t)}(Y)] = 0 \quad \text{for} \quad t \neq 2$$

It follows that the matrix $\text{cov}_0(h, \partial \log f/\partial\beta)$ required in M has possible nonzero entries in the first p elements of its first p rows, has possible nonzero entries in the last $p(p+1)/2$ elements of its next $p(p+1)/2$ rows, and zero elements everywhere else. Hence, M is the direct sum of a $p \times p$ matrix (corresponding to the $L_s^{(1)}(y)$), a $\{p(p+1)/2\} \times \{p(p+1)/2\}$ matrix (corresponding to the $L_s^{(2)}(y)$), and the order $k - p(p+3)/2$ unit matrix.

The score statistic (Equation 6.1.4) involves a vector $\hat{V}$ say, with elements $\hat{V}_r = \sum_j h_r(X_j; \hat{\beta})/\sqrt{n}$. We shall now show that the $\hat{V}_r$ corresponding to the $L_s^{(1)}(y)$ and to the $L_s^{(2)}(y)$ are zero. Given a random sample of size n from the distribution hypothesized to be multivariate normal, we first write Y_j for the random variable corresponding to the jth observation on Y, and Y_{ij} for its ith component. Similarly, we write X_j for the random variable corresponding to the jth observation on X, and X_{ij} for its ith component. Of course, the maximum likelihood estimators for μ and Σ are $\bar{X} = \sum_j X_j/n$ and $\hat{\Sigma} = \sum_j (X_j - \bar{X})(X_j - \bar{X})^T/n$, respectively. We now have

$$\sum_j \hat{Y}_j = \sum_j \hat{\Sigma}^{-1/2}(X_j - \hat{\mu}) = \hat{\Sigma}^{-1/2} \sum_j (X_j - \bar{X}) = 0$$

and

$$\sum_j \hat{Y}_{ij} = 0, \quad i = 1, \ldots, p$$

Also

$$\sum_j \hat{Y}_j\hat{Y}_j^T = \hat{\Sigma}^{-1/2} \sum_j (X_j - \hat{\mu})(X_j - \hat{\mu})^T\hat{\Sigma}^{-1/2} = n\hat{\Sigma}^{-1/2}\hat{\Sigma}\hat{\Sigma}^{-1/2} = nI_p$$

so $\sum_j \hat{Y}_{mj}\hat{Y}_{nj} = n\delta_{mn}$. This is sufficient to show that the terms of degree 1 and 2 are zero because the first degree component contains terms such as $\sum_j H_1(\hat{Y}_{ij}) = \sum_j \hat{Y}_{ij} = 0$, and the second degree component contains terms such as $\sum_j H_2(\hat{Y}_{ij}) = \sum_j (\hat{Y}_{ij}^2 - 1) = 0$ and $\sum_j H_1(\hat{Y}_{mj})H_1(\hat{Y}_{nj}) = \sum_j \hat{Y}_{mj}\hat{Y}_{nj} = 0$ for $m \neq n$. It now follows that

$$S(\hat{\beta}) = \sum_i \hat{V}_i^2, \quad \text{where} \quad \hat{V}_i = \sum_{j=1}^n L_i(\hat{Y}_j)/\sqrt{n}$$

The i summation is over $k - p(p+3)/2$ summands. Since the $\hat{V}_i$ are independent standard normal asymptotically, it is reasonable to group all the $\hat{V}_i$ involving $L_r(\hat{Y})$ of the same degree. So define

$$\hat{U}_s^2 = \sum_i \left\{\sum_j L_i^{(s)}(\hat{Y}_j)\right\}^2 \Big/ n$$

which has the asymptotic χ_v^2 distribution, where $v = {}^{s+p-1}C_s$. It is reasonable to call $\hat{U}_s$ the sth component of $S(\hat{\beta})$. We now have

$$S(\hat{\beta}) = \hat{U}_3^2 + \ldots + \hat{U}_r^2$$

if only orthonormal polynomials of degree at most r are used. Koziol (1986) took $r = 4$. See the discussion in Koziol (1986, pp. 2774–2776) on the use of the components $\hat{U}_s^2$ and their corresponding sub-components, the $\hat{V}_r^2$ (for $r = 3$ and 4). Note that there are ${}^{p+4}C_5$ subcomponents of degree 5, which is 6 for $p = 2$, 21 for $p = 3$ and 56 for $p = 4$. For moderate p the calculation, and then the information to assimilate, is perhaps daunting!

We will show how to construct components. First, write $m_{abcd} = \sum_j \hat{y}_{qj}^a \hat{y}_{rj}^b \hat{y}_{sj}^c \hat{y}_{tj}^d / n$ etc. Note that by the preceding arguments, terms like $m_{10\ldots0}$ and $m_{110\ldots0} (q \neq r)$ are zero, and terms like $m_{110\ldots0} (q = r)$ are 1. Subsequently, zeros are suppressed in this notation. Second, we use $H_1(z) = z$, $H_2(z) = (z^2 - 1)/\sqrt{2}$, $H_3(z) = (z^3 - 3z)/\sqrt{6}$, $H_4(z) = (z^4 - 6z^2 + 3)/\sqrt{24}$, and $H_5(z) = (z^5 - 10z^3 + 15z)/\sqrt{120}$. To obtain the components the orthonormal functions are evaluated at $\beta = \hat{\beta}$ and summed over the sample values. Now for q, r, s, t, u different integers in the range $1, \ldots, p$, the orthonormal functions of degree 5 and the corresponding components are:

$H_5(y_q)$ leading to	$(m_5 - 10m_3)/\sqrt{120}$,
$H_4(y_q)H_1(y_r) = (y_q^4 - 6y_q^2 + 3)/\sqrt{24}y_r$ leading to	$(m_{41} - 6m_{21})/\sqrt{24}$
$H_3(y_q)H_2(y_r) = (y_q^3 - 3y_q)/\sqrt{6}(y_r - 1)/\sqrt{2}$ leading to	$(m_{32} - m_3 + m_{12})/\sqrt{12}$
$H_3(y_q)H_1(y_r)H_1(y_s) = (y_q^3 - 3y_q)/\sqrt{6}y_r y_s$ leading to	$(m_{311} - 3m_{111})/\sqrt{6}$
$H_2(y_q)H_1(y_r)H_1(y_s)H_1(y_t) = (y_q^2 - 1)/\sqrt{2}y_r y_s y_t$ leading to	$(m_{2111} - m_{111})/\sqrt{2}$
and $H_1(y_q)H_1(y_r)H_1(y_s)H_1(y_t)H_1(y_u) = y_q y_r y_s y_t y_u$ leading to	m_{11111}

There are ${}^pC_1$ terms of the form $(m_5 - 10m_3)/\sqrt{120}$, ${}^pC_2$ terms of the form $(m_{41} - 6m_{21})/\sqrt{24}$, and so on, up to ${}^pC_5$ terms of the form m_{11111}. Naturally, if p is 3, for example, then terms requiring more than 3 subscripts cannot appear.

We will now concentrate solely on the bivariate case. The bivariate normal distribution was fundamental in the development of linear regression and correlation (for example, see Galton, 1888). Also, the bivariate normal provides an adequate description of some sets of paired measurements, such as the horizontal and vertical co-ordinates of bullets hitting a target, percentage of fat and protein measurements of milk samples, and heights of fathers and daughters. A well-known graphic check of bivariate normality is to see if there is an elliptical concentration of points in the standard scatterplot. In many situations, however, a scatterplot will be inconclusive and a formal test of significance will give a more objective guide as to the suitability of the bivariate normal distribution. The same remark applies, particularly with small samples, to the graphic checks based on radii and angles that are discussed, for example, by Gnanadesikan (1977, Chapter 5).

Suppose $(x_{11}, x_{12}), (x_{21}, x_{22}), \ldots, (x_{n1}, x_{n2})$ is a bivariate random sample of size n. The statistics $\hat{U}_3^2$ and $\hat{U}_4^2$ for the bivariate case are defined in terms of (y_{i1}, y_{i2}), $i = 1, 2, \ldots, n$ where

$$\begin{pmatrix} y_{i1} \\ y_{i2} \end{pmatrix} = \begin{pmatrix} \hat{\sigma}_1^{-1} & 0 \\ -r/\{\hat{\sigma}_1\sqrt{(1-r^2)}\} & 1/\{\hat{\sigma}_1\sqrt{(1-r^2)}\} \end{pmatrix} \begin{pmatrix} x_{i1} - \hat{\mu}_1 \\ x_{i2} - \hat{\mu}_2 \end{pmatrix}, \; i = 1, 2, \ldots, n$$

in which

$$\hat{\mu}_j = n^{-1} \sum_{i=1}^{n} x_{ij}, \qquad \hat{\sigma}_j^2 = n^{-1} \sum_{i=1}^{n} (x_{ij} - \hat{\mu}_j)^2, \qquad j = 1, 2$$

and

$$r = n^{-1} \sum_{i=1}^{n} (x_{i1} - \hat{\mu}_1)(x_{i2} - \hat{\mu}_2)\hat{\sigma}_1^{-1}\hat{\sigma}_2^{-1}$$

The estimates $\hat{\mu}_j$, $\hat{\sigma}_j^2$ and r of course, are just the usual maximum likelihood estimates. The matrix of the transformation satisfies $A\,\Sigma\,A^T = I_2$. We repeat that this X to Y transformation is not unique, and some, such as Pettitt (1979), give an alternative to our choice. We use

$$y_{i1} = (x_{i1} - \hat{\mu}_1)\hat{\sigma}_1^{-1}, \qquad y_{i2} = \{(x_{i2} - \hat{\mu}_2)\hat{\sigma}_2^{-1} - r(x_{i1} - \hat{\mu}_1)\hat{\sigma}_1^{-1}\}(1 - r^2)^{-1/2}$$

$i = 1, 2, \ldots, n$ because we know of no other transformation that is as simple arithmetically. Using the method demonstrated previously, we find

$$\hat{U}_3^2 = n\{(m_{21}^2 + m_{12}^2)/2 + (m_{30}^2 + m_{03}^2)/6\}$$

and

$$\hat{U}_4^2 = n\{(m_{22} - 1)^2/4 + (m_{31}^2 + m_{13}^2)/6 + [(m_{04} - 3)^2 + (m_{40} - 3)^2]/24\}$$

where $m_{rs} = n^{-1} \sum_{i=1}^{n} \hat{y}_{i1}^r \hat{y}_{i2}^s$. The subcomponents of $\hat{U}_3^2$ are $\hat{V}_{31} = m_{21}/\sqrt{2}$, $\hat{V}_{32} = m_{12}/\sqrt{2}$, $\hat{V}_{33} = m_{03}/\sqrt{6}$, and $\hat{V}_{34} = m_{30}/\sqrt{6}$, while those of $\hat{U}_4^2$ are $\hat{V}_{41} = (m_{22} - 1)/2$, $\hat{V}_{42} = m_{31}/\sqrt{6}$, $\hat{V}_{43} = m_{13}/\sqrt{6}$, $\hat{V}_{44} = (m_{04} - 3)/\sqrt{24}$ and $\hat{V}_{45} = (m_{40} - 3)/\sqrt{24}$. The squared subcomponents are all asymptotically distributed as χ_1^2, so that $\hat{U}_3^2$, $\hat{U}_4^2$ and the omnibus combination $(\hat{U}_3^2 + \hat{U}_4^2)$ are asymptotically distributed as χ_4^2, χ_5^2, and χ_9^2, respectively.

$\hat{U}_3^2$ and $\hat{U}_4^2$ are generalizations of the first two nonzero components of Lancaster's test for univariate normality discussed in §6.2 and by Best and Rayner (1985b). Lancaster's univariate test is defined in terms of Hermite orthogonal polynomials and $\hat{U}_3^2$ and $\hat{U}_4^2$ can similarly be defined in terms of pairwise products of these polynomials. Furthermore, just as the components of Lancaster's univariate test are terms of the Gram–Charlier Type A series, the subcomponents of $\hat{U}_3^2$ and $\hat{U}_4^2$ are terms of the bivariate Type AA series (for example, see Elderton and Johnson, 1969, p. 146).

$\hat{U}_3^2$ and $\hat{U}_4^2$ are also similar to components of a large sample test of bivariate normality for grouped data suggested by Lancaster (1958). Our components are not the same as Lancaster's; he uses a different standardization. Bera and John (1983) proposed tests of multivariate normality based on subsets of $\hat{U}_3^2$ and $\hat{U}_4^2$. Their test statistics, however, depend on a particular X to Y transformation; hence, they are not invariant. Also, from Equation 2.22 of Mardia (1970), or otherwise, it is easy to see that $\hat{U}_3^2 = nb_{1,2}/6$ where $b_{1,2}$ is Mardia's skewness statistic.

To investigate the small sample null distribution of these statistics, 5,000 random bivariate normal samples of size n, where $n = 10$, 12, 14, 16, 18, 20, 25, 30 (10) 100, and 200, were generated. Using standard regression techniques, 5% approximate critical values for $\hat{U}_3^2$, $\hat{U}_4^2$, and $(\hat{U}_3^2 + \hat{U}_4^2)$ were found to be $9.488(1-3.5/n)$, $11.070(1-6.2/n)$, and $16.919(1-6.9/n + 1.1/\sqrt{n})$. These should be used for $10 \le n \le 40$, while χ^2 critical values for $n > 40$ are adequate. Sizes of tests based on these critical values for $\alpha = 0.05$ were also checked using Monte Carlo samples (of size 1,000) and the errors were less than 1%. For $10 \le n \le 40$, the approximate 5% critical values for $\hat{U}_3^2$ agree well with those given in Table 2 of Mardia (1974). For larger n the differences are not important in terms of test size. Approximate P values, if needed, can easily be found via simulation.

It should be noted that the values of the subcomponents of $\hat{U}_3^2$ and $\hat{U}_4^2$ may vary with different X to Y transformations. One invariant subset of unsquared subcomponents of $\hat{U}_4^2$ is Mardia's kurtosis measure $b_{2,2} = m_{40} + m_{04} + 2m_{22}$.

Power Comparisons

We now give approximate powers for $\hat{U}_3^2$, $\hat{U}_4^2$, $(\hat{U}_3^2 + \hat{U}_4^2)$, $b_{2,2}$, and S_W^2. This last statistic was recommended by Mardia (1986) and was defined in Mardia and Foster (1983, pp. 212–213); $b_{2,2}$ is included because we were interested in comparing $b_{2,2}$ and $\hat{U}_4^2$. We used the same alternatives as did Malkovich and Afifi (1973), so the powers we give can also be compared with those of the statistics presented in their Figures A and B. So the alterntives we used for (X_1, X_2) were:

1. (X_1, X_2) independent $\log(N(0, 1))$ variables,
2. (X_1, X_2) independent $U(0, 1)$ variables,
3. (X_1, X_2) independent t_4 variables,
4. mixed bivariate normal samples with 12 pairs independent $N(0, 1)$ variables and 38 pairs independent $N(3, 1)$ variables,
5. mixed bivariate normal samples with 25 pairs independent $N(0, 1)$ variables and 25 pairs independent $N(0, 3)$ variables,
6. mixed bivariate normal samples with 25 pairs independent $N(0, 1)$ variables and 25 bivariate normal pairs with means zero, variances unity and correlation 0.9,
7. mixed bivariate normal samples with 12 pairs independent $N(0, 1)$

Table 6.10 Power comparisons of tests of bivariate normality, $\alpha = 0.05$

Alternative	n	$\hat{U}_3^2$	$\hat{U}_4^2$	$b_{2,2}$	$(\hat{U}_3^2 + \hat{U}_4^2)$	S_W^2
(1)	10	0.57	0.49	0.40	0.55	0.34
(2)	25	0.00	0.05	0.44	0.03	0.32
(3)	10	0.18	0.17	0.12	0.18	0.10
(4)	50	0.28	0.01	0.04	0.05	0.05
(5)	50	0.18	0.23	0.20	0.23	0.16
(6)	50	0.20	0.39	0.22	0.37	0.21
(7)	50	0.97	0.31	0.17	0.80	0.85

variables and 38 bivariate normal pairs with means of three, variances of three and correlation 0.9.

Table 6.10 shows powers based on 500 Monte Carlo samples of size n for these alternatives, $\alpha = 0.05$ and various n. Clearly, no one statistic dominates. Mardia's kurtosis statistic, $b_{2,2}$, is the best for alternative (2), while, as might have been expected, $\hat{U}_3^2$ is best for alternative (4). The omnibus statistic $(\hat{U}_3^2 + \hat{U}_4^2)$ does well for alternatives with longer tails than the normal and for alternatives with correlations. Which of $\hat{U}_4^2$ or $b_{2,2}$ is the better measure of kurtosis? We suggest $\hat{U}_4^2$ is better because it has greater power than $b_{2,2}$ for correlated alternatives and because if has the added flexibility of conveniently defined and distributed subcomponents. The power of $\hat{U}_4^2$ for alternative (2) is disappointing, but see Example 6.6.3.

Example 6.6.1

Quality of a milk sample is assessed partly in terms of its fat and protein content. In New South Wales, various milk factories or depots measure fat and protein of milk from the local suppliers (farmers). One of the tasks of the N.S.W. Dairy Corportion, a state government statutory authority, is to check on the accuracy of these fat and protein measurements. Previous study has shown that if a homogeneous milk sample is subdivided into portions that are measured for fat and protein by each of the factories, then the measurements are independently normally distributed with standard deviations 0.04 and 0.05, respectively.

Figure 6.6 gives data for 26 factories from measurements on one such milk sample. The numbers on the plot indicate factories. The elliptical curve is a 99% confidence band based on the normality assumptions just stated and indicates that factories numbered 14 and 23 gave fat readings that were too high. In fact, these factories used a less precise measurement technique. We use $(\hat{U}_3^2 + \hat{U}_4^2)$ to investigate the bivariate normality of this data, which

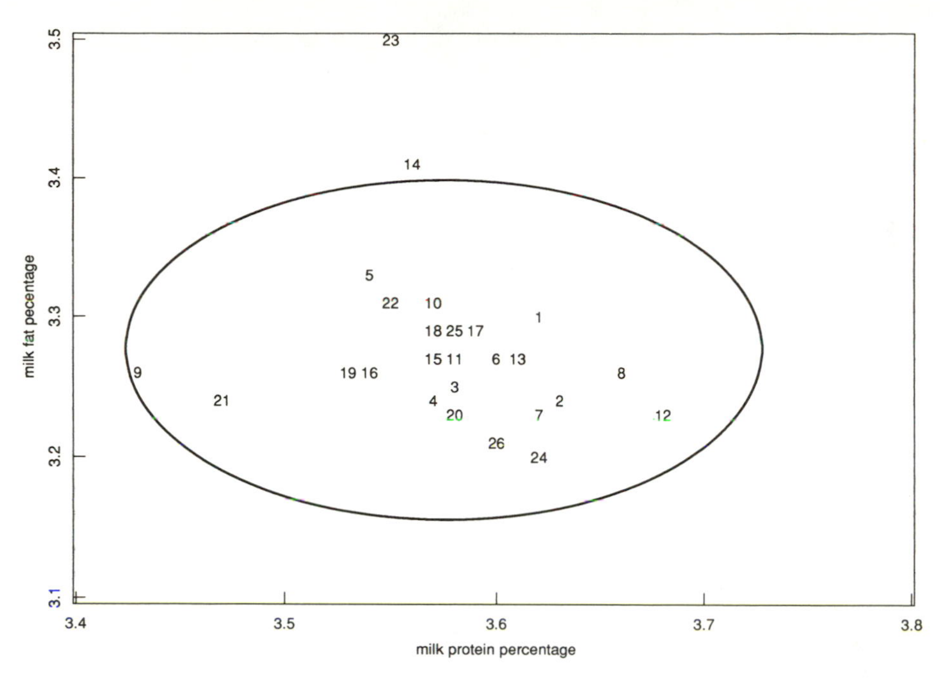

Figure 6.6 Fat and protein readings.

is as follows:

(3.30, 3.62), (3.24, 3.63), (3.25, 3.58), (3.24, 3.57), (3.33, 3.54), (3.27, 3.60), (3.23, 3.62), (3.26, 3.66), (3.26, 3.43), (3.31, 3.57), (3.27, 3.57), (3.23, 3.68), (3.27, 3.61), (3.40, 3.56), (3.27, 3.57), (3.26, 3.54), (3.29, 3.59), (3.29, 3.57), (3.26, 3.53), (3.23, 3.58), (3.24, 3.47), (3.31, 3.55), (3.50, 3.55), (3.20, 3.62), (3.29, 3.57), (3.21, 3.60).

The value obtained is $(\hat{U}_3^2 + \hat{U}_4^2) = 64.0$, which is highly significant. In particular, the components relating to the marginal skewness and kurtosis of the fat measurements are particularly large. Recall that the standardization we use does not affect the first variable. As sample moments are particularly sensitive to outliers, which we know is a problem from Figure 6.6, $(\hat{U}_3^2 + \hat{U}_4^2)$ was recalculated with the two large fat observations removed. This gives $(\hat{U}_3^2 + \hat{U}_4^2) = 12.0$, and the subcomponents are 0.20, −2.12, 0.52, −1.05, −0.54, 1.80, −0.40, 0.48, −1.48. The value of $(\hat{U}_3^2 + \hat{U}_4^2)$ is now insignificant and one or two of the subcomponents do not account for most of its value. We could therefore summarize the reduced data as being bivariate normal with mean vector $(3.27, 3.58)^T$, standard deviations 0.033 and 0.053, and correlation −0.22. The standard deviations are in good agreement with the values 0.04, 0.05 suggested by previous studies.

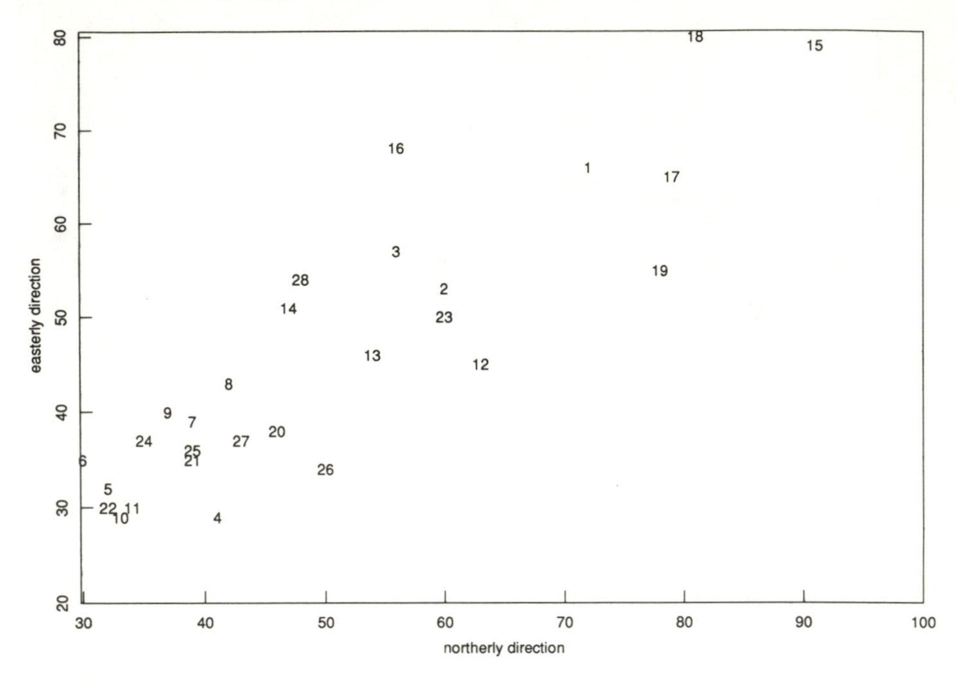

Figure 6.7 Bark thickness on cork trees.

Example 6.6.2

Rao (1948) gave data for the bark thickness on 28 cork trees for four directions on each tree. The N and E observations were:

(72, 66), (60, 53), (56, 57), (41, 29), (32, 32), (30, 35), (39, 39), (42, 43), (37, 40), (33, 29), (32, 30), (63, 45), (54, 46), (47, 51), (91, 79), (56, 68), (79, 65), (81, 80), (78, 55), (46, 38), (39, 35), (32, 30), (60, 50), (35, 37), (39, 36), (50, 34), (43, 37), (48, 54).

Figure 6.7 is a scatterplot of the data where the numbers indicate tree number. Pearson (1956) suggested a trend with tree number and clustering of consecutive numbers is certainly suggested by our figure also.

By testing for bivariate normality, we find $(\hat{U}_3^2 + \hat{U}_4^2) = 6.13$ with subcomponents 1.75, 0.62, 0.04, 1.49, −0.38, 0.24, 0.19, 0.03, −0.47. Although $(\hat{U}_3^2 + \hat{U}_4^2) = 6.13$ and is not significantly large, $\{(m_{30}^2)/6 + (m_{12}^2)/2\}/(\hat{U}_3^2 + \hat{U}_4^2) = 0.86$, so that just two of the subcomponents account for 86% of its value. Both of these subcomponents relate to skewness, the first subcomponent in particular identifies significant marginal skewness in the N observations. Thus, we might suspect that this skewness is important. It shows in only two subcomponents, its effect in $\hat{U}_3^2$ (or $b_{1,2}$) and $(\hat{U}_3^2 + \hat{U}_4^2)$ being diluted by other smaller subcomponents. Pettitt (1979), using other tests, also suggested these data were not bivariate normal. Neither of Mardia's statistics $b_{1,2}$ or $b_{2,2}$, however, is significant.

Example 6.6.3

The statistic $\hat{U}_4^2$ did not perform well for alternative (2)-type data in the power comparisons of the previous section. Consider the following 20 bivariate data points:

(53, 25), (1, 19), (70, 64), (73, 82), (73, 82), (48, 62), (91, 74), (20, 29), (34, 92), (91, 24), (87, 61), (15, 3), (3, 91), (78, 22), (78, 48), (62, 64), (6, 94), (15, 63), (20, 15), (42, 7).

These are derived from random $U(0, 100)$ samples. For this data, $\hat{U}_4^2 = 3.67$, which is nowhere near the 5% critical value; however, the subcomponents of $\hat{U}_4^2$ are -1.36, -1.22, -0.27, -0.47, 0.22, and $\{(m_{40}-3)^2+(m_{04}-3)^2\}/(24\hat{U}_4^2) = 0.91$. As in the previous example, this indicates that two possibly significant components have had their effect diluted.

6.7 Components of the Rao–Robson X^2 Statistic

We will now use a modification of Theorem 6.1.2 to find the components of the Rao–Robson X^2 statistic, X_{RR}^2, which was discussed in §2.4. This section concentrates on mathematical details and may be omitted on first reading by those so inclined.

The general form of X_{RR}^2 and its components will be given subsequently. The components of X_{PF}^2 are found in §7.2. A comparison of the X^2 tests for composite null hypotheses based on X_{PF}^2 and X_{RR}^2 is undertaken in §7.4. For reasons that will become apparent later, we have not simulated the small sample size and power of the components.

Suppose X is a continuous random variable whose probability density function is $f(x; \beta)$, which depends on a vector of nuisance parameters $\beta = (\beta_1, \ldots, \beta_q)^T$. Define an *order k* probability density function by

$$g_k(x; \theta, \beta) = C(\theta, \beta) \exp\left\{\sum_{i=1}^{k} \theta_i h_i(x, \beta)\right\} f(x, \beta)$$

Note that $\theta = (\theta_i)^T$ and $h = (h_i(x; \beta))$ are $k \times 1$. Subsequently, the zero subscript means expectation is taken with respect to the null probability density function $f(x; \beta)$. The following theorem is proved in the same way that Theorems 6.1.1 and 6.1.2 were, and merely permits nonorthogonal $\{h_i(x, \beta)\}$.

Theorem 6.7.1. Assume a random sample $X_1, \ldots, X_n$ is taken from the distribution with probability density function $g_k(x; \theta, \beta)$. Define φ by $\theta = B\varphi$, where B is a $k \times p$ matrix of elements $b_{ij} = b_{ij}(\beta)$ that depend on β; $\hat{B} = (b_{ij}(\hat{\beta}))$. The score statistic for testing $\varphi = 0$ against $\varphi \neq 0$, assuming all derivatives and expectations exist, is

$$S(\hat{\beta}) = \hat{U}^T \hat{B}(\hat{B}^T \hat{M} \hat{B}^T)^{-1} \hat{B}^T \hat{U}$$

provided $\hat{B}^T\hat{M}\hat{B}$ is nonsingular, where

$$\hat{U} = U(\hat{\beta}) = \left\{\left(\sum_j h(X_j; \hat{\beta})\right) - E_0[h(\hat{\beta})]\right\} \Big/ \sqrt{n}, \quad \hat{M} = M(\hat{\beta})$$

and

$$M(\hat{\beta}) = (\text{cov}_0(h, h^T)) - \text{cov}_0\left(h, \frac{\partial \log f}{\partial \beta}\right)$$

$$\times \left\{\text{cov}_0\left(\frac{\partial \log f}{\partial \beta}, \frac{\partial \log f}{\partial \beta}\right)\right\}^{-1} \text{cov}_0\left(\frac{\partial \log f}{\partial \beta}, h\right)$$

To derive X^2_{RR} assume that a continuous random variable is observed, and the maximum likelihood estimators of the nuisance parameters are based on ungrouped data. Then the data are grouped into m categories using boundaries $-\infty = c_0 < c_1 < \ldots < c_m = \infty$. The $\{h_i(x; \beta)\}$ may be taken to be the indicator functions so that $h_i(x; \beta) = 1$ if $x \in (c_{i-1}, c_i]$, $= 0$ otherwise. To conform to Rao and Robson (1974), only the first $k = (m-1)$ h_i are taken, so U is $(m-1) \times 1$. Put $p = m - 1$ also. Then B and $C_h = \text{cov}_0(h, h^T)$ are $(m-1) \times (m-1)$. We assume that the $(m-1) \times q$ matrix $W = \text{cov}_0(h, \partial \log f / \partial \beta)$ is of rank q and the $q \times q$ matrix $C_\beta = \text{cov}_0(\partial \log f / \partial \beta, \partial \log f / \partial \beta)$ is of rank q. Then, from Theorem 6.7.1,

$$S(\beta) = U^T B\{B^T(C_h - WC_\beta^{-1}W^T)B\}^{-1}B^T U$$

Now if B is taken to be I_{m-1}, then $S(\hat{\beta})$ becomes the Rao–Robson statistic defined by Rao–Robson (1974, p. 1144).

Now, write $\hat{C}_\beta$ and $\hat{W}$ for C_β and W with β replaced by $\hat{\beta}$. Since $C_h - \hat{W}\hat{C}_\beta^{-1}\hat{W}^T$ is real symmetric and positive semidefinite, as shown by Anderson (1958, p. 341) there exists an $\hat{F}$ such that $\hat{F}^T(C_h - \hat{W}\hat{C}_\beta^{-1}\hat{W}^T)\hat{F} = I_{k-1}$. Put $B = \hat{F}$ in Theorem 6.7.1. Then, $S(\hat{\beta}) = \hat{V}_1^2 + \ldots + \hat{V}_{k-1}^2 = \hat{V}^T\hat{V}$, and the $\hat{V}_i$ are the required orthogonal components of X^2_{RR}.

7

Neyman Smooth Tests for Categorized Composite Null Hypotheses

7.1 Neyman Smooth Tests for Composite Multinomials

We will now imitate the approach used in §5.2 to construct a class of score tests for smooth alternatives to a multinomial model that depends on unspecified or nuisance parameters. From this class, we will extract the Pearson–Fisher X^2 test and its components. This extends the results of Best and Rayner (1987a), reported in §5.3, on the Pearson X^2 test. These results may be compared with the Rao–Robson X^2 test and its components as discussed in §6.7. Also extracted are (1) an extension to the composite case of the overlapping cells X^2 tests, discussed in §5.6, and (2) cell focusing tests that can assess *contrasts* between cells.

As an alternative to the null hypothesis that m cell probabilities are $p_j = p_j(\beta)$, $i = 1, \ldots, m$, take the order k "smooth" probability function

$$\pi_j(\beta) = C(\theta, \beta) \exp\left\{\sum_{i=1}^{k} \theta_i h_{ij}(\beta)\right\} p_j(\beta), \qquad j = 1, \ldots, m \qquad (7.1.1)$$

Here $\beta = (\beta_1, \ldots, \beta_q)^T$ is a vector of nuisance parameters, $\theta = (\theta_1, \ldots, \theta_k)^T$ is a vector of real parameters and $C(\theta, \beta)$ is a normalizing constant ensuring that $\sum_j \pi_j(\beta) = 1$. For each i, $i = 1, \ldots, k$, the $h_{ij}(\beta)$ are values taken by a random variable H_i with $P(H_i = h_{ij}(\beta)) = \pi_j(\beta)$, $j = 1, \ldots, m$. Since the parameter space has dimension $m - 1$ we must have $k \leq m - 1$. We wish to test for the $p_j(\beta)$ by testing $H: \theta = 0$ against $K: \theta \neq 0$.

As shown previously, a random sample of n observations is taken and N_j, the number of observations in the jth cell, $j = 1, \ldots, m$, is noted. Write n_j for a realization of N_j, and note that $n = \sum_j n_j = \sum_j N_j$. Subsequently, the argument β is dropped from $p_j(\beta)$, $\pi_j(\beta)$ and $h_{ij}(\beta)$, which are assumed to be continuously differentiable. Expectation with respect to the distribution with probability function π_j is denoted by E_k; expectation with respect to the distribution with probability function p_j is denoted by E_0. Write $H = (h_{ij})$ and $N = (N_j)$. Several derivatives will be required. These will be listed without proof in two lemmas. All expectations are assumed to exist.

Lemma 7.1.1. Suppose P is a random variable that takes the value p_j with probability π_j, $j = 1, \ldots, m$ and H_r is a random variable that takes the value h_{rj} with probability π_j, $j = 1, \ldots, m$. Then

$$\frac{\partial \pi_j}{\partial \theta_r} = \pi_j\left(\frac{\partial \log C}{\partial \theta_r} + h_{rj}\right) \tag{7.1.2}$$

$$\frac{\partial \pi_j}{\partial \beta_u} = \pi_j\left\{\frac{\partial \log C}{\partial \beta_u} + \frac{\partial \log p_j}{\partial \beta_u} + \sum_i \left(\theta_i \frac{\partial h_{ij}}{\partial \beta_u}\right)\right\} \tag{7.1.3}$$

$$\frac{\partial \log C}{\partial \theta_r} = -E_k[H_r] \tag{7.1.4}$$

$$\frac{\partial^2 \log C}{\partial \theta_r \, \partial \theta_s} = -\text{cov}_k(H_r, H_s) \tag{7.1.5}$$

$$\frac{\partial \log C}{\partial \beta_u} = -E_k\left[\frac{\partial \log P}{\partial \beta_u}\right] - \sum_i \left(\theta_i E_k\left[\frac{\partial H_i}{\partial \beta_u}\right]\right) \tag{7.1.6}$$

$$\frac{\partial^2 \log C}{\partial \beta_u \, \partial \beta_v} = -E_k\left[\frac{\partial^2 \log C}{\partial \beta_u \, \partial \beta_v}\right] - \sum_i \left(\theta_i E_k\left[\frac{\partial^2 H_i}{\partial \beta_u \, \partial \beta_v}\right]\right) \tag{7.1.7}$$

Suppose now that $G = G(X, \beta)$ is a random variable whose derivatives with respect to θ_r, $r = 1, \ldots, k$, and β_u, $u = 1, \ldots, q$ exist and have finite expectations. The subsequent derivatives also follow routinely, and are also presented without proof.

Lemma 7.1.2. Provided the expectations and derivatives concerned all exist,

$$\frac{\partial E_k[G]}{\partial \theta_r} = \text{cov}_k(G, H_r) \tag{7.1.8}$$

$$\frac{\partial E_k[G]}{\partial \beta_u} = E_k\left[\frac{\partial G}{\partial \beta_u}\right] + \sum_i \left\{\theta_i \,\text{cov}_k\left(G, \frac{\partial H_i}{\partial \beta_u}\right)\right\} + \text{cov}_k\left(G, \frac{\partial \log P}{\partial \beta_u}\right) \tag{7.1.9}$$

The required score statistic may now be derived.

Theorem 7.1.1. Suppose that $\text{cov}_0(\partial \log P/\partial \beta, H)$ is a $q \times k$ matrix with typical element

$$\sum_j \frac{\partial \log p_j}{\partial \beta_u}(h_{rj} - E_0[H_r])p_j = \sum_j \frac{\partial p_j}{\partial \beta_u}(h_{rj} - E_0[H_r])$$

that $\hat{\beta}$ is the maximum likelihood estimator of β and that $\hat{\Sigma} = \Sigma(\hat{\beta})$, in which $\Sigma = \Sigma(\beta) =$

$$\text{cov}_0(H, H) - \text{cov}_0\left(H, \frac{\partial \log P}{\partial \beta}\right)\left\{\text{cov}_0\left(\frac{\partial \log P}{\partial \beta}, \frac{\partial \log P}{\partial \beta}\right)\right\}^{-1} \text{cov}_0\left(\frac{\partial \log P}{\partial \beta}, H\right)$$

Then the score statistic for testing $H_0: \theta = 0$ against $K: \theta \neq 0$ with Equation 7.1.1 is

$$\hat{S}_k = (N - n\hat{p})^T \hat{H}^T \hat{\Sigma}^{-1} \hat{H}(N - n\hat{p})/n \tag{7.1.10}$$

where $\hat{p} = (p_i(\hat{\beta}))$ and $\hat{H} = (h_{ij}(\hat{\beta}))$.

Proof. The logarithm of the likelihood is

$$l = \text{constant} + \sum_{j=1}^{m} n_j \ln \pi_j$$

so that

$$\frac{\partial l}{\partial \theta_r} = \sum_j \frac{n_j}{\pi_j} \frac{\partial \pi_j}{\partial \theta_r} = \sum_j n_j(h_{rj} - E_k[H_r])$$

using Equations 7.1.2 and 7.1.4. It follows that the efficient score is given by

$$U_\theta = (\partial l/\partial \theta_r)\,|_{\theta=0} = HN - nHp$$

for $E_0[H_r] = \sum_j h_{rj} p_j = (Hp)_r$. Similarly from Equations 7.1.3 and 7.1.6,

$$\frac{\partial l}{\partial \beta_u} = \sum_j n_j \left(\frac{\partial \log p_j}{\partial \beta_u} - E_k\left[\frac{\partial \log P}{\partial \beta_u}\right] + \sum_i \theta_i \left\{ \frac{\partial h_{ij}}{\partial \beta_u} - E_k\left[\frac{\partial H_i}{\partial \beta_u}\right]\right\}\right).$$

These are the maximum likelihood equations for the β_u under the unrestricted model. Now

$$\frac{\partial^2 l}{\partial \theta_r\, \partial \theta_s} = -n \frac{\partial E_k[H_r]}{\partial \theta_s} = -n \operatorname{cov}_k(H_r, H_s)$$

using Equation 7.1.8,

$$\frac{\partial^2 l}{\partial \theta_r\, \partial \beta_u} = \sum n_j \frac{\partial h_{rj}}{\partial \beta_u} - n\left\{ E_k\left[\frac{\partial H_r}{\partial \beta_u}\right] + \sum \theta_i \operatorname{cov}_k\left(H_r, \frac{\partial H_i}{\partial \beta_u}\right) + \operatorname{cov}_k\left(H_r, \frac{\partial \log P}{\partial \beta_u}\right)\right\}$$

using Equation 7.1.9, and

$$\frac{\partial^2 l}{\partial \beta_u\, \partial \beta_v} = \sum_j \frac{\partial^2 \log P}{\partial \beta_u\, \partial \beta_v} - E_k\left[\frac{\partial^2 \log P}{\partial \beta_u\, \partial \beta_v}\right] + \operatorname{cov}_k\left(\frac{\partial \log P}{\partial \beta_u}, \frac{\partial \log P}{\partial \beta_v}\right)$$
$$+ \text{ terms with factor } \sum_i \theta_i$$

again using Equation 7.1.9. The expectations of these derivatives are taken, evaluated at $\theta = 0$, and inserted in the appropriate matrices. It follows that

$$I_{\theta\theta} = (-E_0[\partial^2 l/\partial \theta_r\, \partial \theta_s]) = n(\operatorname{cov}_0(H_r, H_s))$$

$$I_{\theta\beta} = (-E_0[\partial^2 l/\partial \theta_r\, \partial \beta_u]) = n\left(\operatorname{cov}_0\left(H_r, \frac{\partial \log P}{\partial \beta_u}\right)\right)$$

$$I_{\beta\beta} = (-E_0[\partial^2 l/\partial \beta_u\, \partial \beta_v]) = n\left(\operatorname{cov}_0\left(\frac{\partial \log P}{\partial \beta_u}, \frac{\partial \log P}{\partial \beta_v}\right)\right)$$

The asymptotic covariance matrix of $H(N - np)/\sqrt{n}$ is $\Sigma = I_{\theta\theta} - I_{\theta\beta} I_{\beta\beta}^{-1} I_{\beta\theta}$. Substitution gives the theorem.

Corollary 7.1.1. The asymptotic covariance matrix of $H(N-np)/\sqrt{n}$ is

$$\Sigma = H\{D - pp^T - W^T(WD^{-1}W^T)^{-1}W)H^T \qquad (7.1.11)$$

where $D = \text{diag}(p_1, \ldots, p_m)$ and $W = (W_{uj})$, in which $W_{uj} = \partial p_j/\partial\beta_u$. The matrix Σ has rank $k \le m-1-q$, provided the $k \times m$ matrix H has rank at least k.

Proof. We only need to express the $I_{\theta\theta}$, $I_{\theta\beta}$, and $I_{\beta\beta}$ in matrix form:

$$\text{cov}_0(H_r, H_s) = \sum_j h_{rj}h_{s}p_j - \left(\sum_j h_{rj}p_j\right)\left(\sum_j h_{sj}p_j\right) = (HDH^T)_{rs} - (Hp)_r(Hp)_s;$$

$$\text{cov}_0(H_r, \partial \log P/\partial\beta_u) = \sum_j h_{rj}(\partial \log P/\partial\beta_u)p_j$$

$$= \sum_j h_{rj}(\partial p_j/\partial\beta_u) = \sum_j h_{rj}W_{uj} = (HW^T)_{ru}$$

using

$$E_0[\partial \log P/\partial\beta_u] = \sum_j (\partial \log p_j/\partial\beta_u)p_j = \sum_j \partial p_j/\partial\beta_u = \partial 1/\beta_u = 0;$$

$$\text{cov}_0(\partial \log P/\partial\beta_u, \partial \log P/\partial\beta_v) = \sum_j (\partial p_j/\partial\beta_u)p_j^{-1}(\partial p_j/\partial\beta_v)$$

$$= \sum_j W_{uj}p_j^{-1}W_{vj} = (WD^{-1}W^T)_{uv}$$

To obtain the rank of Σ note that

$$\Sigma = HD^{1/2}(I_m - K)D^{1/2}H^T$$

where $K = D^{-1/2}pp^TD^{-1/2} + D^{-1/2}W^T(WD^{-1}W^T)^{-1}WD^{-1/2}$, I_m is the $m \times m$ unit matrix and $p = (p_j)$. Using the fact that $\Sigma_j W_{uj}p_j^{-1}p_j = \Sigma_j W_{uj} = \Sigma_j \partial p_j/\partial\beta_u = \partial 1/\partial\beta_u = 0$, $WD^{-1}p = 0$, and it is easily seen that K is idempotent. The rank of $I_m - K$ is therefore equal to its trace, which is $m-1-q$.

Corollary 7.1.2. Under the null hypothesis, the maximum likelihood equations for the nuisance parameters $\beta_1, \ldots, \beta_q$ are

$$WD^{-1}N = 0$$

Proof. Under the null hypothesis the maximum likelihood equations are

$$\frac{\partial l}{\partial\beta_u} = \sum_j n_j \frac{\partial \log p_j}{\partial\beta_u} = \sum_j n_j p_j^{-1}\frac{\partial p_j}{\partial\beta_u} = (WD^{-1}N)_u = 0$$

One implication of Corollary 7.1.1 is that the asymptotic distribution of $\hat{S}_k$ is χ_k^2, central under the null hypothesis $H_0: \theta = 0$ and noncentral under contiguous alternatives. Note that Σ is more conveniently calculated using Equation 7.1.11 than the form given in the theorem.

In order to access the components of the score statistic we need a modification of Theorem 7.1.1.

Theorem 7.1.2. Define φ by $\theta = B\varphi$ where B is a $k \times p$ matrix of elements $b_{ij} = b_{ij}(\beta)$ that depend on β. The score statistic for testing $H_0: \varphi = 0$ against $K: \varphi \neq 0$ with the regular cases of Equation 7.1.1, is

$$\hat{S}_k = (N - n\hat{p})^T \hat{H}^T \hat{B} (\hat{B}^T \hat{\Sigma} \hat{B})^{-1} \hat{B}^T \hat{H} (\hat{N} - n\hat{p})/n \tag{7.1.12}$$

provided $\hat{B}^T \hat{\Sigma} \hat{B}$ is nonsingular.

Proof. The efficient score is $U_\varphi = B^T U_\theta$, and the information is $I_\varphi = B^T I_\theta B$ (for example, see Cox and Hinkley, 1974, p. 130). The theorem now follows from Theorem 7.1.1.

Since $\hat{\Sigma}$ is a positive semidefinite real symmetric matrix, there exists a matrix F such that $F^T \hat{\Sigma} F = I_k$ (again, see Anderson 1958, p. 341). Choose $\hat{B} = F$, so that the matrix of the quadratic form $\hat{S}_k$ is nonsingular. In addition, however, the components are now accessible because if $\hat{V} = \hat{B}^T \hat{H}(N - n\hat{p})/\sqrt{n}$, then $\hat{S}_k = \hat{V}^T \hat{V} = \sum_i \hat{V}_i^2$, where $\hat{V} = (\hat{V}_i)$. The $\hat{V}_i$ are jointly asymptotically multivariate normal with covariance matrix I_k, and are thus asymptotically independent.

7.2 Components of the Pearson–Fisher Statistic

The well-known Pearson–Fisher X^2 statistic was discussed in §2.4. It is defined by

$$X_{PF}^2 = \sum_{j=1}^{m} (N_j - n\hat{p}_j)^2/(n\hat{p}_j)$$

where $\hat{p}_1, \ldots, \hat{p}_m$ are the maximum likelihood estimators of $p_1, \ldots, p_m$, respectively, using the categorized data. Cox and Hinkley (1974, p. 326) showed how to derive X_{PF}^2 as a score statistic for an appropriate model. We show that X_{PF}^2 is a particular case of $\hat{S}_{m-q-1}$ from §7.1, and obtain components by choosing $\hat{H}$ appropriately.

We choose $\hat{H}$ so that $\hat{\Sigma} = I_{m-q-1}$. Then

$$\hat{S}_{m-q-1} = (N - n\hat{p})^T \hat{H}^T \hat{H} (N - n\hat{p})/n.$$

This is achieved if the $\hat{h}_{rs}$ satisfy several constraints. First, for r and $s = 1, \ldots, m-q$ and m, we require

$$\sum_{j=1}^{m} \hat{h}_{rj} \hat{h}_{sj} \hat{p}_j = \delta_{rs}, \quad \text{with} \quad \hat{h}_{mj} = 1, j = 1, \ldots, m$$

It follows that

$$\sum_{j=1}^{m} \hat{h}_{rj} \hat{p}_j = \delta_{rm} = (\hat{H}\hat{p})_r = 0, \quad \text{for} \quad r = 1, \ldots, m-q-1$$

Thus, in matrix terms

$$\hat{H}\hat{D}\hat{H}^T = I_{m-q-1} \text{ and } \hat{H}\hat{p} = 0$$

If we further insist that

$$\hat{H}\hat{W}^T = 0$$

then $\hat{\Sigma} = I_{m-q-1}$ as required.

Now define $\hat{H}^{*T} = [\hat{H} \mid \hat{D}^{-1}\hat{p} \mid \hat{D}^{-1}\hat{W}^T(\hat{W}\hat{D}^{-1}\hat{W}^T)^{-1/2}]$. That $\hat{H}^*\hat{D}\hat{H}^{*T} = I_m$ follows after noting that $\hat{p}^T\hat{D}^{-1}\hat{p} = 1$ and $\hat{W}\hat{D}^{-1}\hat{p} = 0$. As a result of this construction, $\hat{D}^{-1} = \hat{H}^{*T}\hat{H}^*$, from which

$$\hat{H}^T\hat{H} = \hat{D}^{-1} - \hat{D}^{-1}\hat{p}\hat{p}^T\hat{D}^{-1} - \hat{D}^{-1}\hat{W}^T(\hat{W}\hat{D}^{-1}\hat{W}^T)^{-1}\hat{W}\hat{D}^{-1}$$

By substituting in $\hat{S}_{m-q-1}$, we obtain

$$\hat{S}_{m-q-1} = (N - n\hat{p})^T\hat{D}^{-1}(N - n\hat{p})/n$$

as $\hat{p}^T\hat{D}^{-1}(N - n\hat{p}) = 0$ and $\hat{W}\hat{D}^{-1}(N - n\hat{p}) = 0$.

The components of X^2_{PF} are now obvious. For $\hat{H}$ as earlier, define $\hat{V} = \hat{H}(N - n\hat{p})/\sqrt{n} = \hat{H}N/\sqrt{n}$. Then $X^2_{PF} = \hat{V}^T\hat{V}$ and the elements $\hat{V}_i$ of $\hat{V}$ are asymptotically independent χ^2_1 random variables. As in the simple case, it can be shown that $\hat{V}_i$ optimally detects θ_i and no other θ_j because it can be shown that $\hat{V}_i$ is a score statistic for an appropriate model. The elements $\hat{V}_i$ of $\hat{V}$ are asymptotically orthonormal components of X^2_{PF} and give a breakdown of the deviations from the nominated distribution. Each $\hat{V}_i^2$ has asymptotic distribution χ^2_1, and is asymptotically independent of the other components. If $\hat{H}$ is based on a Helmert matrix, the ith component gives a comparison between the first i classes and the $(i + 1)$th class.

X^2_{PF} requires the grouped maximum likelihood estimators, which are usually more difficult to calculate than the corresponding ungrouped estimators. If the ungrouped data are available, however, an ungrouped goodness of fit test not based on the multinomial would be used. The major difficulty with the components of X^2_{PF} is constructing $\hat{B}$ conveniently so that the components are meaningful. The same may be said of the components of X^2_{RR} found in §6.7. An alternative to using these computationally difficult components is given in the next section. Another approach is to use the double-root residuals (DRRs) discussed in §6.4.

7.3 Composite Overlapping Cells and Cell-focusing X^2 Tests

For testing a given hypothesis, it may be known that a particular parametric test is more powerful than its nonparametric competitors. Assume that the data are categorized. To verify the applicability of the parametric test we would like to check the distributional assumptions. If in fact the null hypothesis is composite, the Pearson–Fisher test is suggested. We will focus on two problems.

Perhaps only a small data set is available because observations are in some sense expensive. Alternatively, we may be particularly interested in assessing if the tails follow the specified distribution; the extreme 10% of the tails may be important in subsequent hypothesis testing. In either case, some cell expectations will be small, and the asymptotic χ^2 distribution of the test statistic cannot be relied upon. In §5.6, using a technique introduced by Hall (1985), we suggested overlapping the cells. Our simulations there demonstrated the success of overlapping in achieving actual test sizes closer to the nominal sizes when compared with nonoverlapping cells tests. We are interested in extending this investigation to the composite situation.

At the outset it is worth recording that this concern with the asymptotic distribution disappears if a P value can be simulated. The cell-focusing techniques developed here, however, are of interest even if this is the case.

In this section, we will first extract some overlapping cells tests from Theorems 7.1.1 and 7.1.2. In addition, tests focusing on deviations between the observed and expected cell frequencies for particular (sets of) cells can be obtained. A limited simulation study demonstrates that a technique involving isolating cells looks distinctly promising, but fails to demonstrate the efficacy of overlapping in the Poisson case we will consider.

The cell-focusing tests may be constructed from Theorems 7.1.1 and 7.1.2 as follows. If each row of H has a 1 in only one position and zeros otherwise, then $H(N - np)$ is composed of the observed minus expected for the corresponding cells. Then $\hat{S}_k$ assesses the adequacy of the fit of the hypothesized distribution in the indicated cells. For example, if

$$H_1 = \begin{pmatrix} 0 & 0 & 0 & 1 & 0 & 0 & 0 & 0 & \dots \\ 0 & 0 & 0 & 0 & 1 & 0 & 0 & 0 & \dots \\ 0 & 0 & 0 & 0 & 0 & 1 & 0 & 0 & \dots \\ 0 & 0 & 0 & 0 & 0 & 0 & 1 & 0 & \dots \end{pmatrix}$$

then $\hat{S}_4$ focuses on cells 4 to 7.

A composite overlapping cells X^2 tests results if we take

$$H_2 = \begin{pmatrix} 1 & 1 & 1 & 1 & 0 & 0 & 0 & 0 & \dots \\ 1 & 1 & 1 & 0 & 1 & 0 & 0 & 0 & \dots \\ 1 & 1 & 1 & 0 & 0 & 1 & 0 & 0 & \dots \\ 1 & 1 & 1 & 0 & 0 & 0 & 1 & 0 & \dots \end{pmatrix}$$

Cells 4 to 7 have each been overlapped with the first three cells. Then $H_2(N - np)$ will consist of observed minus expected for each of the cells 4 to 7 amalgamated with cells 1, 2, and 3. In the Pearson X^2 test, higher cell expectations result in a better χ^2 approximation to the null distribution of the test statistic. The aim here is to increase the cell expectations from np_i to $n(Hp)_i$, in the hope that the χ^2 approximation to the null distribution of the overlapped statistic will be an improvement on that of the nonoverlapped statistic.

We will now present simulations for a Poisson null only, and for H_1 and H_2 as given earlier, but these are suggestive. One feature of the Poisson is that it has infinitely many cells and so allows us to demonstrate how to handle this situation.

Suppose a random sample of size n is taken from a Poisson distribution with mean μ, so that the probability function is

$$p_j(\mu) = e^{-\mu}\mu^j/j!, \qquad j = 0, 1, 2, \ldots$$

In the constructions that follow, the experimenter assumes that the Poisson mean μ is in the vicinity of 1.0. This assumption enables appropriate cells to be overlapped, namely those with relatively high expectation. In some cases this could be achieved by an appropriate choice of units. The alternative is to overlap many cells, and this does not seem desirable for demonstration purposes. Subsequently, the jth cell corresponds to the event the Poisson random variable takes the value $j-1$.

Calculation of Σ requires matrix multiplication of a $k \times m$ matrix. But m is infinite in the Poisson case. This causes no great difficulty in practice, except in explicitly writing the test statistic. To do this, first note that $(WD^{-1}W^T)^{-1} = \mu$. Then write $Hp = p^*$ and $HW^T = q^*$. Now

$$(\Sigma_{ij}) = \sum_r h_{ir}p_r h_{jr} - (p^*)_i(p^*)_j - \mu(q^*)_i(q^*)_j \tag{7.3.1}$$

in which

$$(p^*)_i = \sum_r h_{ir}p_r$$

and

$$(q^*)_i = \sum_r h_{ir}(r/\mu - 1)p_r$$

If we wish to focus on just the $i+1$th cell, H becomes a vector with just one row, with a 1 in the $i+1$th position and zeros elswhere. From Theorem 7.1.1 the appropriate statistic is $\hat{T}_i$, where

$$\hat{T}_i = (N_i - n\hat{p}_i)^2/\{n[\hat{p}_i - \hat{p}_i^2 - \hat{\mu}(i/\hat{\mu} - 1)^2 p_i^2]\}. \tag{7.3.2}$$

If just two cells were to be focused upon, then H has 1's in the $(1, i+1)$th and $(2, j+1)$th positions and zeros elswehere. From Equations 7.1.10 and 7.1.11, the score statistic is $\hat{T}_{ij}$, where, if $a = (i/\hat{\mu} - 1)$ and $b = (j/\hat{\mu} - 1)$, then

$$\hat{T}_{ij} = \frac{\{\hat{p}_j - \hat{p}_j^2(1+b^2)\}(N_i - n\hat{p}_i)^2 + 2(1+ab)\hat{p}_i\hat{p}_j(N_i - n\hat{p}_i)(N_j - n\hat{p}_j) + \{\hat{p}_i - \hat{p}_i^2(1+a^2)\}(N_j - n\hat{p}_j)^2}{\{n\hat{p}_i\hat{p}_j[1 - \hat{p}_i(1+a^2) - \hat{p}_j(1+b^2) + (a-b)^2\hat{p}_i\hat{p}_j]\}} \tag{7.3.3}$$

In general, we prefer the matrix representation of the $\hat{S}_k$ using Equations 7.1.10 and 7.1.11. Theorem 7.1.2 permits their orthogonal components to be calculated. Thus, for both H_1 and H_2 we obtain $\hat{S}_{(i)}$, $i = 2$, 3, and 4 and their components $\hat{V}_i$, $i = 1, \ldots, 4$.

To see the effect of overlapping, the following simulation study was undertaken. For $\alpha = 0.05$ and several values of n and μ, a sample of size n

was drawn from a pseudorandom Poisson distribution with mean μ. The statistics $\hat{S}_{(i)}$, $i = 2$, 3, and 4 their components, $\hat{V}_1$, $\hat{V}_2$, $\hat{V}_3$, and $\hat{V}_4$ (defined in §7.1), were then calculated. This experiment was repeated 5,000 times during which the component values that exceeded the $100\alpha\%$ point of the χ_1^2 distribution, and the values of $\hat{S}_{(i)}$ that exceeded the $100\alpha\%$ point of the χ_i^2 distribution were counted. Table 7.1 shows some of the results. By any reasonable criterion, when μ is close to 1, the sizes for H_1 are closer to 5% than those for H_2. For larger μ there is little to choose between the H's. There is a slight improvement as n increases. If we focus on just $\hat{V}_3^2$ and $\hat{V}_4^2$, for which the cell expectations are quite small, the sizes corresponding to H_1 are not adequate, but overlapping has not improved the situation. For the other statistics, the H_1 sizes are quite acceptable, and it is not surprising that H_2 has failed to improve upon them. In fact, for other than $\hat{V}_3^2$ and $\hat{V}_4^2$, the sizes are all quite good considering n is not large.

Table 7.1 Simulated sizes multiplied by 10,000 with nominal size 5%; based on 5,000 simulations

n	μ	$\hat{V}_1^2$	$\hat{V}_2^2$	$\hat{V}_3^2$	$\hat{V}_4^2$	$\hat{S}_{(2)}$	$\hat{S}_{(3)}$	$\hat{S}_{(4)}$
				(a) Matrix H_1				
10	0.8	47	41	9	1	46	38	26
	1.0	40	46	21	5	45	44	33
	1.2	35	39	39	11	40	48	37
	2.6	53	43	32	47	41	42	41
	4.0	45	47	42	38	57	52	46
	5.5	35	39	48	52	40	45	44
20	0.8	34	56	18	4	53	47	37
	1.0	31	42	45	9	39	54	49
	1.2	30	41	56	19	35	54	53
	2.6	51	39	38	38	45	41	37
	4.0	51	50	48	39	49	49	45
	5.5	42	51	57	47	43	48	51
				(b) Matrix H_2				
10	0.8	27	14	15	6	29	24	26
	1.0	28	15	20	8	30	29	33
	1.2	31	17	29	11	32	36	38
	2.6	44	49	35	40	43	40	43
	4.0	46	54	51	46	46	46	46
	5.5	45	44	47	41	38	44	44
20	0.8	30	25	22	9	41	42	37
	1.0	34	28	40	12	41	54	49
	1.2	32	36	46	20	39	55	54
	2.6	42	47	45	41	44	41	40
	4.0	48	47	54	47	46	48	45
	5.5	46	51	54	51	47	47	49

The low probability cells used in this study have expectation 0.002, 0.012, 0.077, and 0.383 when $n = 10$ and $\mu = 0.8$; for $n = 10$ and $\mu = 2.6$ these have improved to 0.319, 0.735, 1.414, and 2.176. In the former case, the sizes are not adequate and in the latter they are. In §5.6 it was found that the sizes were improved when overlapping cells with expectations of about 0.1. When cell expectations were 0.5, sizes were usually adequate with or without overlapping. It may be that with a cell expectation below 0.05 even overlapping will not help; for cell expectations above 0.5, there is no need to overlap; and in between it is not likely that overlapping will improve upon an already adequate performance. Of course the numbers 0.05 and 0.5 need further investigation.

Although the study failed to confirm the efficacy of the overlapping technique, the adequacy of the χ^2 approximation to the small sample null distribution corresponding to H_1 is confirmed. Perhaps the same regime concerning cell expectations applies approximately for all members of the $\hat{S}_k$ class.

Example 7.3.1

The number of deaths due to horsekicks in the Prussian army, given in Andrews and Herzberg (1985), is a well-known data set. For the amalgamation of the 14 corps, the number of years in which there were 0, 1, . . . deaths were

Number of deaths	0	1	2	3	4	5	6	7	8	9	10	11	12	13	14	15	16	17	18
Number of years	0	0	0	1	1	2	2	1	1	2	1	3	1	0	1	2	0	1	1

The mean is 9.8 deaths per year. Suppose that before sighting the data we had particular interest in 10 and 15 deaths per year. Using Equations 7.3.2 and 7.3.3 we obtain $\hat{T}_{10} = 1.026$, $\hat{T}_{15} = 3.414$, and $\hat{T}_{10,15} = 4.172$. The two latter statistics are significant at the 10% level, but not the 5% level, casting doubt on the Poisson model.

Using the approach of §6.4 we find $\hat{S}_4 = 8.99$ with components $\hat{V}_2 = 2.76$, $\hat{V}_3 = -0.84$, $\hat{V}_4 = -0.80$, and $\hat{V}_5 = -0.10$. The large $\hat{V}_2$ and the fact that $\hat{V}_2$ accounts for a large proportion of $\hat{S}_4$ confirm the previous conclusion.

One problem with the use of X^2_{PF} is the necessity to use grouped maximum likelihood estimators, which can be awkward to calculate in some circumstances. If the data follow a discrete distribution with infinitely many possibilities, such as the Poisson or negative binomial, then the formulation of this section permits the use of the ungrouped maximum likelihood estimator because m can be taken to be very large. The grouped and ungrouped maximum likelihood estimators, therefore, are for all practical purposes the same. Of course, this option is not available in calculating X^2_{PF}.

For the reader applying the methodology proposed in this section we have a word of warning. Some apparently sensible choices for H yield singular covariance matrices. This is symptomatic of repeating a contrast that has already been made. While we can handle this technically, it seems undesirable in practice. We therefore suggest that if a singular covariance matrix arises, a different choice of H be used.

7.4 A Comparison between the Pearson–Fisher and Rao–Robson X^2 Tests

We will turn now to some comments made by D. S. Moore in D'Agostino and Stephens (1986). He states that "Chi-squared tests are generally less powerful than EDF tests and special purpose tests of fit." This raises several points.

1. The X^2 tests are omnibus tests and should not be compared with directional tests, which is how we interpret "special purpose." Tests based on the components are directional. We suggest that omnibus tests should be compared with omnibus tests, and directional tests with directional tests. Like the smooth tests we advocate, the empirical distribution function (EDF) tests have components (for example, see Stephens, 1974b). It would be interesting to see power comparisons between the smooth and EDF test components, but we have not undertaken this. We merely note that our smooth components, unlike the EDF components, have convenient χ^2 distributions.
2. The X^2 tests are tests of categorized data and can only be compared fairly with other tests of categorized data. The EDF tests are typically tests of continuous distributions. Interesting comparisons would be between the smooth tests of Chapters 4 and 6 and tests based on the empirical distribution function.
3. In some of our early papers we showed that by choosing the number of classes appropriately we could obtain X^2 tests at least as powerful as some of the EDF tests. Inasmuch as one (or possibly more) X^2 test was being chosen from a class of tests, this could be argued as being unfair to the EDF tests. On the other hand, the X^2 tests are score tests and are therefore optimal in the senses described in §3.4 and by Singh (1987) (see the following). Tests that are more powerful for certain alternatives are clearly not subject to the same constraints, and will be correspondingly less powerful for other alternatives.

Perhaps the appropriate conclusion to draw from these points is that in choosing to use one test rather than another we should be aware of the properties of each. For example, if the data are severely rounded, can a test of continuous data be used? Are particular alternatives anticipated, or is protection against all alternatives required?

With regard to their properties, it is worthwhile to quote Singh (1987, p. 3255).

> It is shown that the statistic (X^2_{RR}) . . . is asymptotically optimal for the family of local alternatives corresponding to the asymptotic reduction of the testing problem within the class of tests based on two sets of statistics; one is the set of cell counts and the other one consists of raw data mle . . . of the nuisance parameter . . . . This optimality is analogous to that of . . . (X^2_{PF}) for a different family of local alternatives which corresponds to the asymptotic reduction of the problem within the class of tests based only on the set of cell counts.

The optimality claimed here is uniformly the most powerful invariant for linear hypotheses (see Lehmann 1986, p. 477).

Moore goes on to say that when testing for normality, the (Watson–)Roy and Rao–Robson tests have power gains of up to 40% compared to the Pearson–Fisher test. Recall that in §2.4 we made the observation that the Rao–Robson test is often claimed to be more powerful than the Pearson–Fisher test, in spite of the claim of Moore and Spruill (1975) that neither is always superior.

We now undertake a small simulation study to assess these claims. The simulation study reported in §6.3 for the exponential was extended to include the Pearson–Fisher and Rao–Robson X^2 tests, each with 3 and 10 classes. All powers are based on 1,000 simulations. One would naively *expect* the $\hat{S}_k$ tests to be more powerful than those based on X^2_{RR}, and these would be expected to be more powerful than the tests based on X^2_{PF}. This is because the $\hat{S}_k$ use more information than the X^2_{RR}, which use more information than the X^2_{PF} tests, this being via the uncategorized data.

We claim we could have *constructed* alternatives to achieve any desired ordering of the tests. Recall that X^2_{PF} is the score test corresponding to the probability function $\pi_j(\beta)$ given by (7.1.1) with $k = m - q - 1$, and X^2_{RR} is the score test corresponding to the probability density function $g_k(x; \theta, \beta)$ given by Equation 6.1.1 with $k = m - 1$ and the h_i indicator functions. So X^2_{PF} detects alternatives in an $m - q - 1$ dimensional parameter space $\mathcal{P}_{PF}$ say, while X^2_{RR} detects alternatives in an $m - 1$ dimensional parameter space say, $\mathcal{P}_{RR}$. Which test is the more powerful depends on the projections of the alternative into $\mathcal{P}_{PF}$, $\mathcal{P}_{RR}$, and the q dimensional space $\mathcal{P}_{RR} - \mathcal{P}_{PF}$. If the alternative projects totally into $\mathcal{P}_{PF}$ then X^2_{PF} will be more powerful than X^2_{RR}, for the parameter of noncentrality for the asymptotic χ^2 distribution is not increased at all, and cannot compensate for the extra degrees of freedom. Similarly, if enough of the alternative projects into $\mathcal{P}_{RR} - \mathcal{P}_{PF}$, then X^2_{RR} must be superior.

Rather than construct the answers we might want, we have taken a relatively arbitrary selection of common alternatives. We compare X^2_{PF}, X^2_{RR}, and the smooth tests $\hat{S}_k$ of exponentiality discussed in §6.3.

Table 7.2 fails to confirm the naive expectations from the previous paragraph. For the $\alpha = 0.05$ tests, the χ^2 mixture and shifted exponential alternatives show the Pearson–Fisher tests to be most powerful, and the Rao–Robson tests least powerful of those considered. For the Pareto (3) alternative the categorized tests are not distinguishable, and all are superior

Table 7.2 Power comparisons of tests for exponentiality; $\alpha = 0.05$ and $n = 20$

Alternative	$X^2_{PF}(3)$	$X^2_{PF}(10)$	$X^2_{RR}(3)$	$X^2_{RR}(10)$	$\hat{S}^2_2$	$\hat{S}^2_4$
Weibull (0.8)	0.07	0.11	0.09	0.18	0.24	0.23
Weibull (1.5)	0.28	0.19	0.28	0.19	0.38	0.32
Uniform (0, 2)	0.11	0.45	0.51	0.43	0.67	0.50
Pareto (3)	1.00	0.99	0.97	0.99	0.69	0.93
Shifted-Pareto (3)	0.06	0.12	0.33	0.38	0.49	0.48
Shifted exponential (0.2)	0.26	0.20	0.10	0.14	0.17	0.17
$(\chi^2_{0.5} + \chi^2_4)/2$	0.57	0.75	0.16	0.65	0.40	0.65
$(\chi^2_1 + \chi^2_5)/2$	0.23	0.32	0.07	0.13	0.09	0.21
null	0.06	0.04	0.05	0.05	—	—

to the $\hat{S}_k$ tests. The other alternatives support the naive expectation. We must admit to being inconsistent: the $X^2_{PF}(3)$, $X^2_{RR}(3)$, and $\hat{S}^2_2$ tests all are order 2 tests, the $\hat{S}^2_4$ test is an order 4 test, and the $X^2_{PF}(10)$ and $X^2_{RR}(10)$ tests are order 9 tests.

The simulations here are consistent with Singh (1987), who proposed a pretest to decide between X^2_{PF} and X^2_{RR}. The pretest assessed whether the maximum likelihood estimators using the categorized and uncategorized data were sufficiently different to justify using X^2_{RR} with its extra degrees of freedom.

Our study supports Moore and Spruill's (1975) assertion of no clearly superior test and consequently, the projection into parameter spaces argument. However, we have a further objection. The test statistics are graduated in that the X^2_{PF} require categorized estimators and categorized data; the X^2_{RR} require uncategorized estimators and categorized data; and the $\hat{S}_k$ require uncategorized estimators and uncategorized data. It is difficult to imagine genuine circumstances in which the use of X^2_{RR} is appropriate. Surely data are either categorized or not? To use X^2_{RR}, we must obtain the uncategorized data, calculate the maximum likelihood estimators, categorize the data and then ignore or lose the uncategorized information?! We suggest that if X^2_{RR} is calculable, then the uncategorized data is usually available, and the $\hat{S}_k$ tests of Chapter 6 are available. These tests are at least typically as convenient as X^2_{RR}, and are usually more powerful. We recommend use of the $\hat{S}_k$ tests and their components for uncategorized data, and for categorized data, the X^2_P and X^2_{PF} tests and whatever orthogonal cell-focusing tests are appropriate for the data.

Although we have theoretically obtained the components of X^2_{PF} and X^2_{RR}, we have not pursued the matter. For the reasons preceding, we do not believe the components of X^2_{RR} are of practical interest. The difficulty with the components of X^2_{PF} is that we have not yet developed a scheme that is computationally convenient with components that are practically relevant. This must await further research. The cell-focusing tests, however, may be more relevant.

8
Conclusion

8.1 Introduction

We have now virtually completed the assignment proposed at the beginning of this monograph: an exposition of the one sample smooth goodness of fit tests. In the next section we would like to draw the reader's attention to other areas that the techniques and tests we have discussed may be applied. We will look at density estimation, outlier detection, normality testing of several samples, and residual assessment in regression analysis.

Finally, we will conclude the monograph with a brief review and return to an example highlighting our approach. We also suggested at the outset that in the broad area of smooth goodness of fit there was much research left to be done. We will propose some topics that we would have liked to investigate, had we been given the time!

8.2 Further Applications of Smooth Tests of Fit

As a goodness of fit test, the test for normality discussed in §6.2 has good power properties in small as well as large samples. In Best and Rayner (1985b), we called this test *Lancaster's test of normality*. Here we focus on applications of this test. This is not to say that *any* of the smooth tests may not be applied in the same way. In the topics discussed in the following subsections, the central idea is the use of the components $\hat{V}_r$ of the smooth test statistic to give information about the "distance" from normality. Eubank et al. (1987) used analogous components to derive a variety of test statistics.

8.2.1. Density Estimation

With the recent advent of better printers and graphics packages, it is fairly easy for the statistician to present good data displays for clients. Density estimates or *smoothed histograms* are easily understood by most nonstatisticians and so are an important method of data presentation and exploratory data analysis. In our experience, they are more easily understood than either Q–Q plots or the newer linked-line charts of Shirahata (1987).

One of the older methods of density estimation is to use an orthogonal series, although a problem is in determining the number of terms to use in the estimate (see Diggle and Hall, 1986, for a recent discussion). One such orthogonal series density estimate, say $\hat{f}(x)$, is the Gram–Charlier Type A series,

$$\hat{f}(x) = \varphi(x)\left\{1 + \sum_{r=1}^{\infty} \hat{V}_r H_r(x)/\sqrt{n}\right\}$$

where $\varphi(x) = \exp(-x^2/2)/\sqrt{(2\pi)}$ and $\hat{V}_r$ and $H_r(x)$ are defined in §6.2. Of course $\hat{V}_0 = 1$, and $\hat{V}_1 = \hat{V}_2 = 0$ because the mean and standard deviation are estimated by maximum likelihood estimation. As shown previously, the $\hat{V}_r$ are asymptotically independent standard normal variables. A rough guide to the choice of terms in $\hat{f}(x)$, therefore, is to consider only those $\hat{V}_r$ for which $|\hat{V}_r| > p$, where p is an appropriate percentage point of the standard normal. Alternatively, on the basis of power considerations, as in §6.2, the statistic $S(\hat{\beta})$ with $k = 6$ performs well, and so $\hat{V}_3$, $\hat{V}_4$, $\hat{V}_5$, and $\hat{V}_6$ could be used. We have compared the plots produced by $\hat{f}(x) = \varphi(x)\{1 + \hat{V}_3 H_3(x) + \ldots + \hat{V}_6 H_6(x)\}/\sqrt{n}$ and a kernel density estimation routine using a Gaussian kernel.

Figure 8.1 compares the plots of length of eruption time for the Old Faithful geyser presented in Table 2.2 of Silverman (1986) and in Weisberg (1980). The plots are fairly similar but the calculations needed for $\hat{f}(x)$ are much simpler. The kernel density estimate is based on a window width of 0.7. Both plots indicate bimodality of the data with modes at about 2 minutes and 4 minutes, although this is more pronounced for $\hat{f}(x)$. Figures 2.1 and 2.2 of Silverman (1986) showed that using histograms with different origins can result in different data descriptions. This is not a problem with the Gram–Charlier density $\hat{f}(x) = \varphi(x)\{1 + \hat{V}_3 H_3(x) + \ldots + \hat{V}_6 H_6(x)\}/\sqrt{n}$. It should be stressed, however, that we are not advocating the use of the Type A series as a panacea for density estimation. Its simplicity and link to Lancaster's test suggest its use as a first attempt at density estimation. Often, the components of Lancaster's test will suggest the Type A series will work better on transformed data. Sometimes, other orthogonal polynomials, such as the Laguerre or Poisson–Charlier, will do better, and sometimes other methods of density estimation will be preferable. An issue we will not discuss here is whether the coefficients $\hat{V}_r$ can be better estimated using robust or bootstrap estimates (see Schemper, 1987); however, we will discuss transformations a little further.

Plotting of the sample value of (β_1, β_2), which follow from $(\hat{V}_3, \hat{V}_4)$, on Figure 2 of Pearson et al. (1977) is often helpful. For example, for the suicide data given in Table 2.1 of Silverman (1986), $(\hat{V}_3, \hat{V}_4) = (8.7, 10.4)$ from which $(\hat{\beta}_1, \hat{\beta}_2) = (5.1, 8.2)$. Then, from Pearson et al. (1977), (5.1, 8.2)

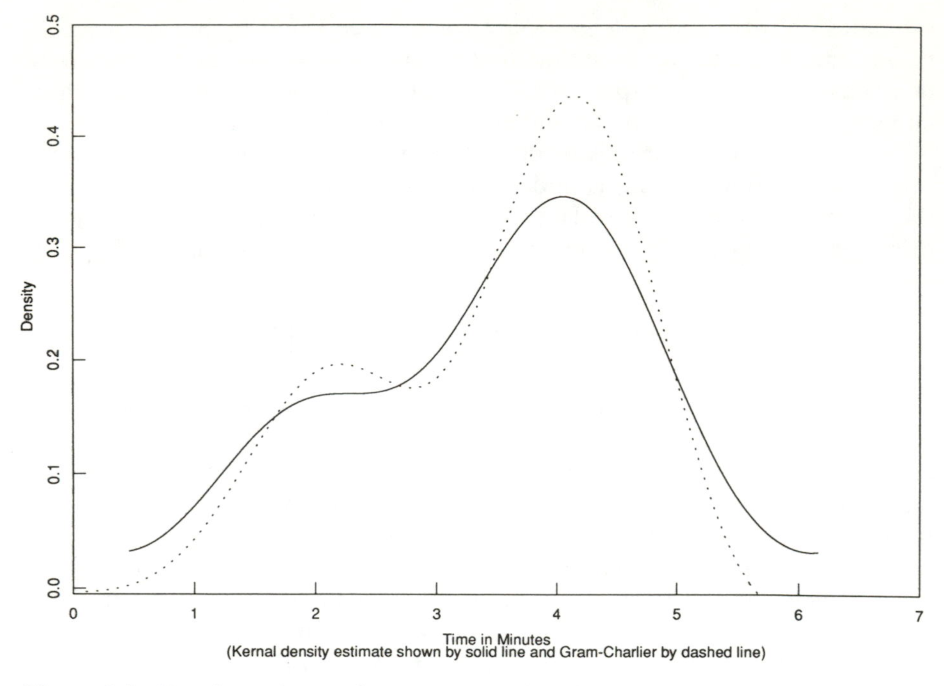

Figure 8.1 Density estimates for geyser eruption times.

is closer to the gamma line than to $(0, 3)$. Hence, a cube root transformation, which makes gamma distributed variables more normal, might be useful. Our Figure 8.2 compares Type A series density estimates for untransformed and cube root transformed data with the kernel density estimate produced as described earlier. Window widths of 105 and 0.6 were used for the raw and cube root data, respectively. Notice that the Type A series is not appropriate for the raw data as it produces negative frequencies for lengths of treatment between about 300 and 420 days. The Type A series and kernel density estimate agree well for the transformed data. A more complicated method of transformation is to find which power of the observations minimizes $S(\hat{\beta})$.

We will now discuss the extension of these ideas to more than one dimension using the extension of the smooth goodness of fit test for the univariate normal to that for the multivariate normal. In particular, we will focus on the tests for bivariate normality discussed in the latter part of §6.6. A bivariate Gram–Charlier series can be given as:

$$\hat{f}(x_1, x_2) = \varphi(x_1)\varphi(x_2)\left\{1 + \sum_{r,s=1}^{\infty} \hat{V}_{rs} H_r(x_1) H_s(x_2)/\sqrt{n}\right\}$$

where the $\hat{V}_{rs}$ are defined in §6.6. As in the univariate case the bivariate Gram–Charlier series can provide a density estimate. Suppose we use the

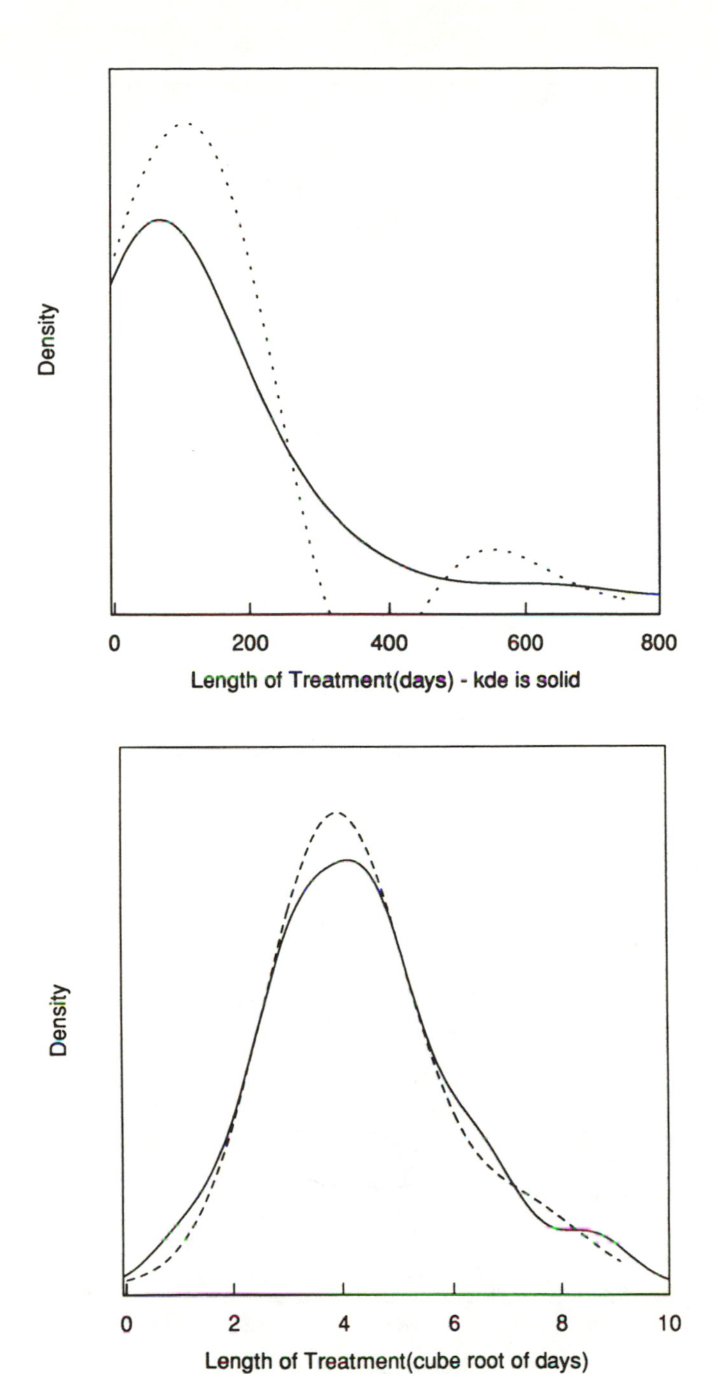

Figure 8.2 Suicide data.

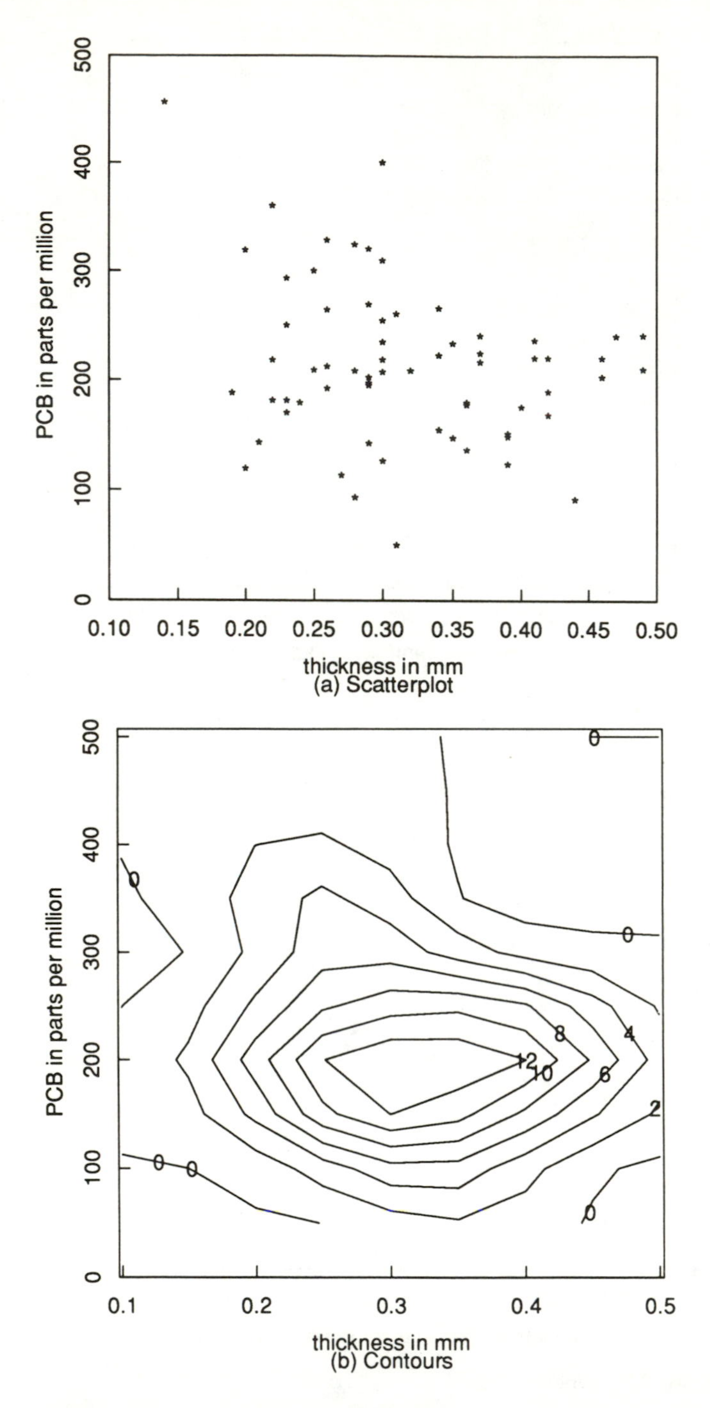

Figure 8.3 Anacapa pelican egg data.

nine nonzero terms with $r+s \leq 4$ to give a density estimate for the following bivariate data taken from Risebrough (1972):

(452, 0.14), (184, 0.19), (115, 0.20), (315, 0.20), (139, 0.21), (177, 0.22), (214, 0.22), (356, 0.22), (166, 0.23), (246, 0.23), (177, 0.23), (289, 0.23), (175, 0.24), (296, 0.25), (205, 0.25), (324, 0.26), (260, 0.26), (188, 0.26), (208, 0.26), (109, 0.27), (204, 0.28), (89, 0.28), (320, 0.28), (265, 0.29), (138, 0.29), (198, 0.29), (191, 0.29), (193, 0.29), (316, 0.29), (122, 0.30), (305, 0.30), (203, 0.30), (396, 0.30), (250, 0.30), (230, 0.30), (214, 0.30), (46, 0.31), (256, 0.31), (204, 0.32), (150, 0.34), (218, 0.34), (261, 0.34), (143, 0.35), (229, 0.35), (173, 0.36), (132, 0.36), (175, 0.36), (236, 0.37), (220, 0.37), (212, 0.37), (119, 0.39), (144, 0.39), (147, 0.39), (171, 0.40), (216, 0.41), (232, 0.41), (216, 0.42), (164, 0.42), (185, 0.42), (87, 0.44), (216, 0.46), (199, 0.46), (236, 0.47), (237, 0.49), (206, 0.49).

For this data, Y_1 = PCB concentration in parts per million for a brown Pelican egg and Y_2 = thickness in millimeters of the same eggshell. The Y_i data were analyzed in Example 1.4.3. The eggs were found at Anacapa and PCB is an industrial pollutant. We find

$$\hat{V}_{30} = 2.31273 \qquad \hat{V}_{40} = 1.98855$$

$$\hat{V}_{03} = 1.25045 \qquad \hat{V}_{04} = -0.742639$$

$$\hat{V}_{21} = -2.48540 \qquad \hat{V}_{31} = -2.196260$$

$$\hat{V}_{12} = -0.482296 \qquad \hat{V}_{13} = 0.571905$$

$$\hat{V}_{22} = -0.528110$$

Also $\bar{y}_1 = 210.14$, $\bar{y}_2 = 0.32$, $\hat{\sigma}_1 = 72.36$, $\hat{\sigma}_2 = 0.08$, and $r = -0.25$. Figure 8.3 gives a scatterplot and contours based on $\hat{f}(x_1, x_2)$ for this data. The magnitudes of the components $\hat{V}_{30}$ and $\hat{V}_{31}$ are larger than expected and indicated the data are not bivariate normal. Furthermore, the Y_1 values are not normally distributed.

The use of contours based on a density estimate to highlight features of a scatterplot appears to be an underused graphic aid. The bivariate Gram–Charlier series provides a simply obtained density estimate. Mardia (1970, paragraph 5 of Chapter 3) warns that the Gram–Charlier approach is unsatisfactory. We suggest that as in the univariate case, perhaps after a preliminary transformation, it is useful as a simple first attempt that is appropriate for data not too far from normality. In practice, the areas of negative density, which seem characteristic of this method, may not be important.

8.2.2. Outlier Detection

The detection of outliers is an important practical problem for which there are many suggested procedures. Barnett and Lewis (1984) summarized many of these but did not advocate any method as being best in most situations. We suppose here that we have a sample from a normal

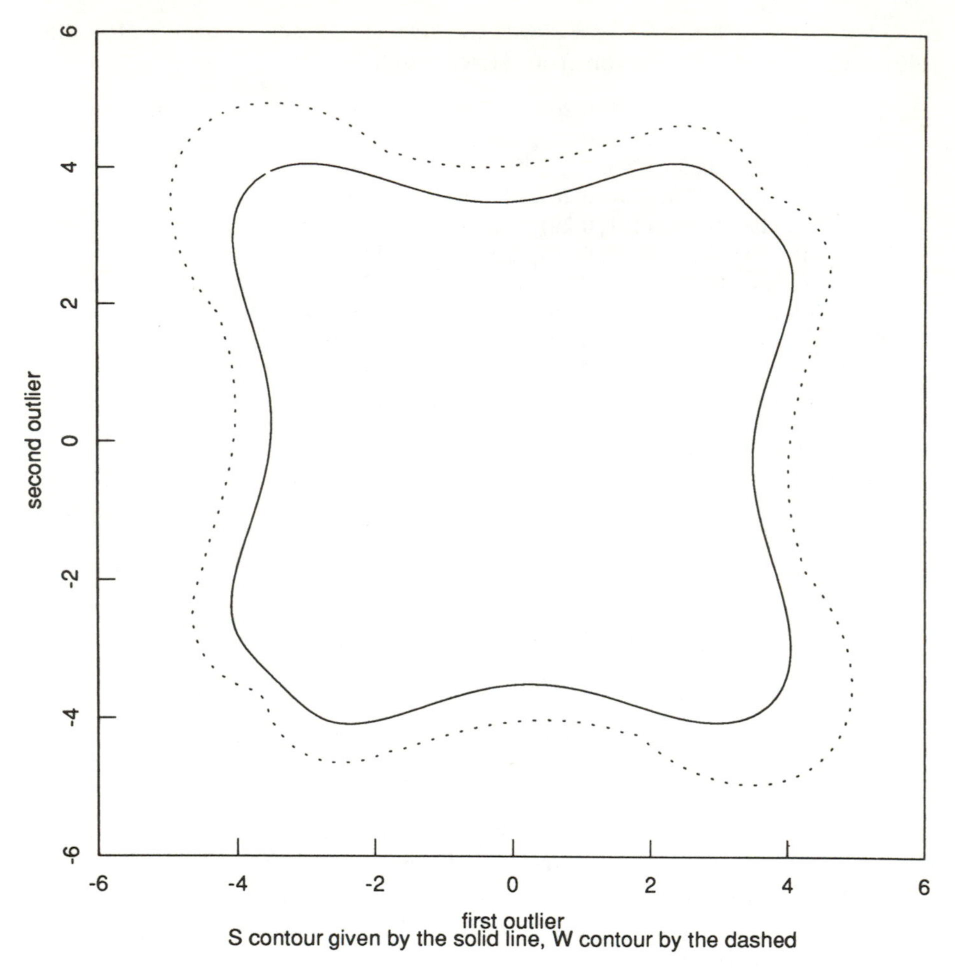

Figure 8.4 Sensitivity contours for W and S, Alpha = 10%, $n = 20$.

population and wish to detect, say, two or three outliers. Barnett and Lewis (1984, p. 175) noted that the skewness and kurtosis coefficients (standardized versions of which are given by $\hat{V}_3$ and $\hat{V}_4$) have certain optimal properties. It is not unreasonable, therefore, to use $S(\hat{\beta})$ with $k = 6$, say $\hat{S}_4$, as a test for outliers. Nonsignificance confirms not only normality, but also the lack of outliers. Figure 8.4 shows a sensitivity plot as suggested by Prescott (1976) for the $\hat{S}_4$ and Shapiro–Wilk W statistics. To construct the curve shown, the following procedure is adopted for each statistic. The values of x_1 and x_2 are found, for example, so that the statistic equals its 10% critical value. This equates the other $(n - 2)$ points in the sample of size n to the expected normal order statistic. Clearly, Lancaster's test statistic, $\hat{S}_4$, outperforms W. The statistic W was compared with $\hat{S}_4$ because

Table 8.1 Approximate P values for $\hat{S}_4$

omit	none	−1.9	−1.9 −0.9	−1.9 −0.9 −0.5
none	.0	.0	.0	.0
5.9	.0	.0	.0	.0
4.7, 5.9	.2	.9	.4	.6
0.4, 4.7, 5.9	.1	.9	.4	.4

W is the only commonly used omnibus normality test that Barnett and Lewis (1984) considered as an outlier test.

Consider checking the following ten observations for two or three outliers. Observations are: −1.9, −0.9, −0.5, −0.4, −0.3, 0.02, 0.2, 0.4, 4.7, and 5.9. Suppose we calculate the approximate P value for the $\hat{S}_4$ statistic using 200 Monte Carlo simulations of random normal samples of size 10. The approximate P value is then just the proportion of these $\hat{S}_4$ values greater than or equal to the originally observed $\hat{S}_4$ value. Do this for zero, one, two, . . . observations deleted. The deleted observations will be successively the i smallest and j largest observations, where i and $j = 0, 1, 2, \ldots$. This produces Table 8.1. This approach is somewhat similar to Kitagawa (1979), except that we use approximate P values rather than an information criterion. Thus, the data sets that produce P values less than say, .05 are to be considered to contain outliers and we seek the largest data set such that $P > .05$, say.

Table 8.1 clearly shows that 4.7 and 5.9 should be regarded as outliers. Dixon's test, as implemented in the 1986 Standards Association of Australia document AS2850, finds no outliers. Dixon's test, like many outlier tests, but not $\hat{S}_4$ and W, is sensitive to "masking" effects.

8.2.3. Normality Testing of Several Samples

In §6.2 and in Best and Rayner (1985b), we illustrated testing normality of single samples and emphasized examining the components, $\hat{V}_r$. Often, however, the applied statistician wishes to test overall normality of several samples. The analysis of variance provides an important example. Table 8.2 shows flavour "scores" from 24 tasters for four tomato varieties: Floradade, Momotaro, Summit, and Rutgers. The "scores" are lengths in millimeters (mm) measured along a line with one end marked "poor" and the other "good." Assume a different set of 24 tasters judged each variety. These results are a very small subset of a large project at the CSIRO Division of Food Research. The project is attempting to improve the flavor of tomatoes.

Approximate Monte Carlo P_i values, for $i = 1, 2, 3,$ and 4, for the "scores" of the four varieties are combined using, for example, Fisher's

Table 8.2 Flavor scores for four tomato varieties

Floradade	Momotaro	Summit	Rutgers
43	74	109	39
5	112	25	82
74	64	48	100
64	101	91	62
10	105	52	126
16	12	35	26
75	33	42	24
20	90	100	35
36	129	22	74
76	37	122	19
60	50	105	113
57	44	119	56
55	18	29	61
29	24	26	21
82	48	102	6
91	62	48	13
66	88	108	118
27	50	53	91
72	73	57	60
108	119	82	88
84	109	105	15
50	50	108	32
82	12	13	134
39	37	74	29

well-known statistic $F = -2\{\ln(P_1) + \ldots + \ln(P_4)\}$, to give an overall P value. For the data of Table 8.2, we find, on the basis of 200 Monte Carlo samples, $P_1 = .37$, $P_2 = .21$, $P_3 = .07$, and $P_4 = .11$. This gives $F = 14.84$, and as this is a χ_8^2 variable, an overall P value of .06 is obtained. It would appear that the normality assumption for this data is doubtful. Examination of the components of $\hat{S}_4$ for the four variety data sets shown in Table 8.2 and inspection of Figure 8.5 indicates that the distribution of the scores are less platykurtic (peaked) than is expected for a normal distribution. Notice the sizes and signs of $\hat{V}_4$ and $\hat{V}_6$ in Table 8.3.

Figure 8.5 shows histograms of flavor scores made by tasters and the Gram–Charlier Type A density estimate truncated at $k = 4$ for each variety. Notice how the density estimates highlight features of the histograms. There is clearly large taster variability possibly caused or confounded with large within variety variability. No variety is preferred. It would be interesting to compare the preceding procedure on the $\hat{S}_4$ statistic with the methods and/or the statistics used by Quesenberry, Giesbrecht, and Burns (1983) and by Pettitt (1977). We will leave this for another time.

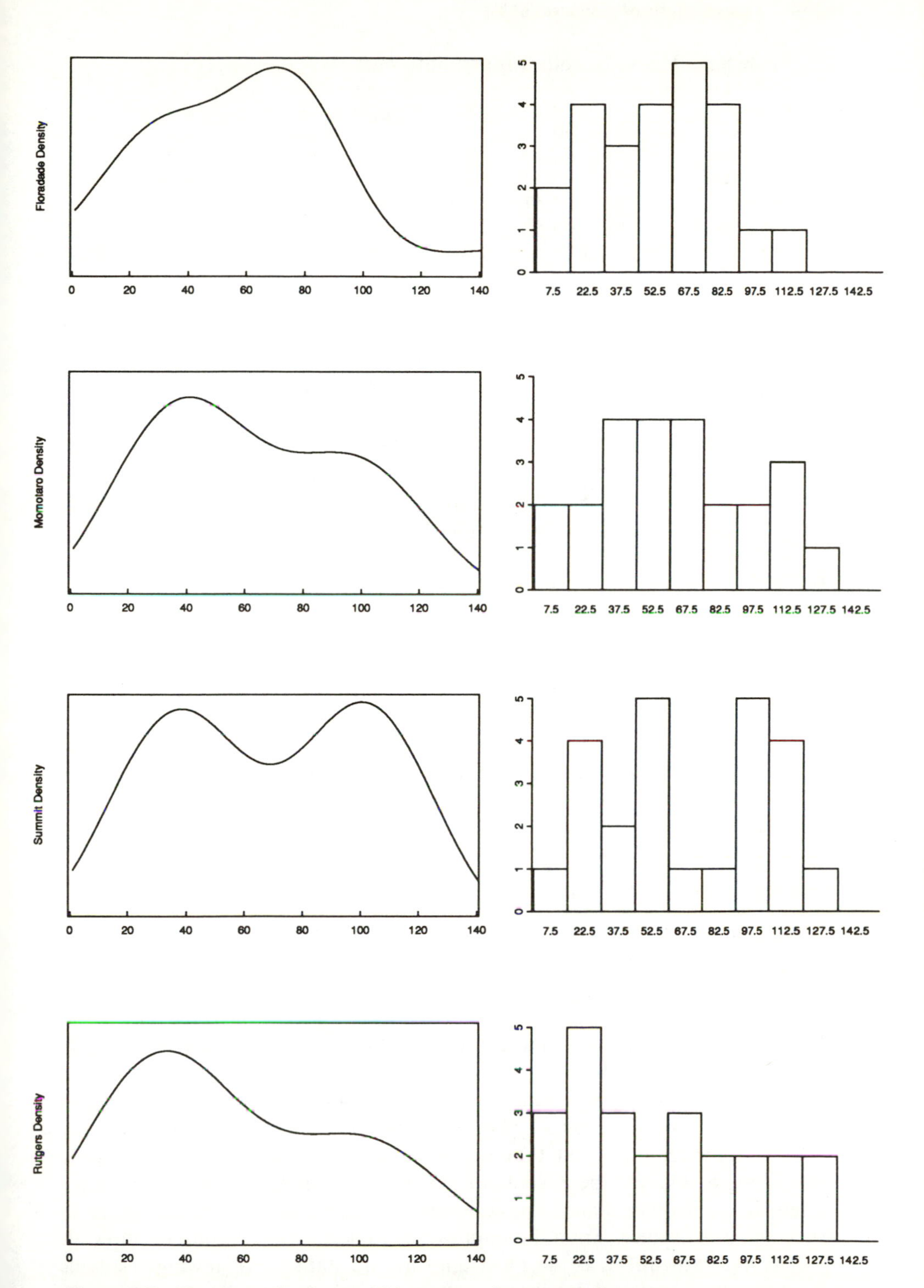

Figure 8.5 Density estimates and histograms for tomato flavor scores.

Table 8.3 $\hat{V}_i$ and $\hat{S}_4$ values for tomato data

Variety	$\hat{V}_3$	$\hat{V}_4$	$\hat{V}_5$	$\hat{V}_6$	$\hat{S}_4$	$(\hat{V}_4^2+\hat{V}_6^2)/\hat{S}_4$
Floradade	−0.34	−0.89	0.59	0.77	1.85	0.75
Momotaro	0.49	−1.11	−0.68	1.10	3.14	0.78
Summit	−0.07	−1.51	0.07	1.87	5.79	0.99
Rutgers	0.86	−1.08	−1.10	1.11	4.35	0.55

8.2.4. Residuals

Suppose we have $Y = X\beta + E$ where Y is an $n \times 1$ vector of observations, X is a fixed $n \times k$ matrix of full rank, β is a $k \times 1$ vector of parameters to be estimated, and E is an $n \times 1$ vector of unobservable stochastic disturbances, assumed to be normally distributed with mean zero and unknown variance σ^2. Given n independent observations on Y, calculate the $\hat{E}_i = Y_i - X\hat{\beta}$, $i = 1, \ldots, n$, where $\hat{\beta}$ is the least squares estimate of β. Then calculate $\hat{S}_4$ for the data $\hat{E}_1, \ldots, \hat{E}_n$. This particular value of $\hat{S}_4$ will now be called $\hat{S}_4^*$.

To find an approximate P value for $\hat{S}_4^*$ we can again use Monte Carlo methods. It is easy to see, for example, that

$$\hat{E} = [I - X(X^TX)^{-1}X^T]E = [I - H]E$$

Thus, to generate random $\hat{E}$ values, we generate n standard normal $N(0, 1)$ values and transform them using the known $I - H$ matrix. We can use $N(0, 1)$ rather than $N(0, \sigma^2)$ values because $\hat{S}_4$ is scale invariant. Thus random $\hat{S}_4$ values, say 200, are obtained for a particular X matrix, and the approximate P value is just the proportion of these equalling or exceeding $\hat{S}_4^*$. Notice that $I - H$ only needs evaluation once.

As an example, consider the 15 residuals

−0.25, −0.26, 0.21, −0.04, −0.14, 0.10, 0.36, −0.09, 0.02, −0.03, 0.27, −0.02, 0.05, −0.01, −0.14

from a simple linear regression with corresponding response variable values

11.0, 11.1, 7.2, 8.3, 12.4, 14.7, 5.1, 21.7, 20.2, 19.1, 25.0, 10.2, 13.3, 23.6, 12.0.

Rayner and Best (1986) ignored the correlations between the 15 residuals and found $\hat{S}_4^* = 1.245$ with an approximate P value of .67. By proceeding as previously, however, we obtain an approximate P value (based on 200 random $\hat{S}_4^*$ values) of 0.63. This is in agreement with Pierce and Gray (1982), who demonstrated that the least squares residuals from a simple linear regression can be tested for normality assuming they are uncorrelated. In cases other than a simple linear regression there is no guarantee of such similar P values. Then the approach outlined earlier could be followed.

It is not known if $\hat{S}_4$ is the most appropriate statistic for testing residuals from linear models for normality. However Jacque and Bera (1987) give some evidence that $\hat{S}_2 = \hat{V}_3^2 + \hat{V}_4^2$ is useful. Based on comparisons between $\hat{S}_2$ and $\hat{S}_4$ in §6.2 and Best and Rayner (1985b), we suggested $\hat{S}_4$ will be competitive with other statistics. Also, note that a test for normality may not suffer "masking" effects, and so may be a useful test for outliers among regression residuals.

A possible difficulty with the method we have just described is that because H is $n \times n$, the method may become unwieldy for large n. Wetherill (1986) suggested using normality tests for uncorrelated data on those realizations of the $\hat{E}_i$ for which $1 - H(i, i) \leq 2(n-k)/(3n)$. Also, because the $\hat{E}_i$ are linear combinations of the E_i, some loss in power is expected when compared to a test of independent identically distributed variables. Huang and Bolch (1974) showed that the ordinary residuals we use do better than BLUS residuals, and an unpublished report of Buckley (1987) showed that recursive residuals are worse than either BLUS or ordinary residuals. We note in passing that Aggarwal and Bajgier (1987) considered that a kurtosis test of regression residuals is useful in constructing an appropriate regression model. As the kurtosis statistic is related to a component of $\hat{S}_4^*$, we expect that $\hat{S}_4^*$ will also be useful for this purpose.

8.3 Concluding Example

Before concluding our investigation into the smooth tests of goodness of fit, we will present one last example. This involves testing for the binomial distribution with parameters N and p. One of the difficulties with the binomial is that the parameter p does not enter the problem as a nuisance parameter. Thus, investigations such as the small sample corrections to the asymptotic null critical points must be repeated for each value of p. In the present example, the sample size is so large that we use the asymptotic critical points. Moreover, we have confidence that the asymptotic optimality properties of the score tests ensure that the smooth tests used are very competitive with whatever other tests are appropriate.

Example 8.3.1

Weldon's dice data were analyzed in Example 1.4.1. The data give the numbers of 5's and 6's in $n = 26{,}306$ throws of $N = 12$ dice. The sample size n is surely sufficiently large that the confirmation of the critical points and optimality are unnecessary. We tested for the binomial distribution with parameters $N = 12$ and $p = 1/3$ using X_P^2. Strong evidence that the distribution is not binomial with parameters $N = 12$ and $p = 1/3$ was found. It would appear from the histogram in Figure 1.1 that the dice are slightly biased, but the analysis so far does not permit that conclusion.

In Theorem 4.2.1, however, the smooth test and its components were found for completely specified uncategorized models. The appropriate orthogonal functions in the binomial case are the Krawtchouk polynomials.

For the binomial probability function

$$f(x, p) = {}^N C_x p^x q^{N-x}, \qquad x = 0, 1, \ldots, N$$

where $q = 1 - p \in (0, 1)$, the normalized Krawtchouk polynomials are defined by $h_0(x) = 1$ and for $r \geq 1$,

$$h_r(x; p) = \left\{ \sum_{u=0}^{N} {}^r C_u \, {}^{N-x} C_{r-u} (q/p)^{u/2} \{-(p/q)\}^{(r-u)/2} \right\} \Big/ \sqrt{{}^N C_r}$$

This gives

$$h_1(x; p) = (x - Np)/\sqrt{(Npq)}$$

and

$$h_2(x; p) = \{(x - Np)^2 + (p - q)x - Npq\}/\{pq\sqrt{[2N(N-1)]}\}$$

The smooth test of order k ($\leq m - 1$) uses the statistic $V_1^2 + \ldots + V_k^2$, where $V_i = \sum_j h_i(X_j; p)/\sqrt{n}$. From the corollary to Theorem 5.1.2 the V_i are also the components of X_P^2.

Now X_P^2 rejects the binomial $N = 12$, $p = 1/3$ hypothesis, and we calculate $V_1 = \sum_j h_1(X_j; p)/\sqrt{n} = 5.20$ to test the hypothesis of a mean (Np) shift, and $V_2 = \sum_j h_2(X_j; p)/\sqrt{n} = 1.54$ to assess a variance shift. Since for this n the V_i can be expected to closely follow their asymptotic $N(0, 1)$ distribution, the mean shift hypothesis is resoundingly confirmed and the variance shift hypothesis rejected. Apparently, biased dice are common in practice. Note that although X_P^2 can reject the distributional hypothesis, it is the component analysis that is informative.

Now suppose we wish to use X_{PF}^2 to test for the binomial distribution with parameter $N = 12$ and with p a nuisance parameter. Pearson (1900) gave $X_{PF}^2 = 17.76$, but we obtain 13.16. Again, the difference is probably due to rounding. The probability p is estimated by maximum likelihood via

$$\hat{p} = \sum_{j=1}^{n} x_j/(Nn) = \sum_{x=0}^{N} x f_x/(Nn)$$

where f_x is the number of observations equal to x, $x = 0, 1, \ldots, N$. These equations for $\hat{p}$ are equivalent to $\hat{V}_1 = \sum_j h_1(X_j; \hat{p})/\sqrt{n} = 0$, and yield $\hat{p} = 0.3377$. We calculate $\hat{V}_2 = \sum_j h_2(X_j; \hat{p})/\sqrt{n} = 0.64$. With 11 degrees of freedom X_{PF}^2 is not significant. Similarly $\hat{V}_2$ finds no evidence against the binomial hypothesis. Notice that $\hat{V}_2$ is a standardized version of R. A. Fisher's binomial index of dispersion,

$$\sum_{x=0}^{N} f_x (x - N\hat{p})^2/(N\hat{p}\hat{q})$$

whereas

$$\hat{V}_2 = \left\{ \sum_{x=0}^{N} f_x (x - N\hat{p})^2/\hat{p}\hat{q} - Nn \right\} \Big/ \sqrt{\{2nN(N-1)\}}$$

In §6.4 we defined double-root residuals (DRRs), and used them in an

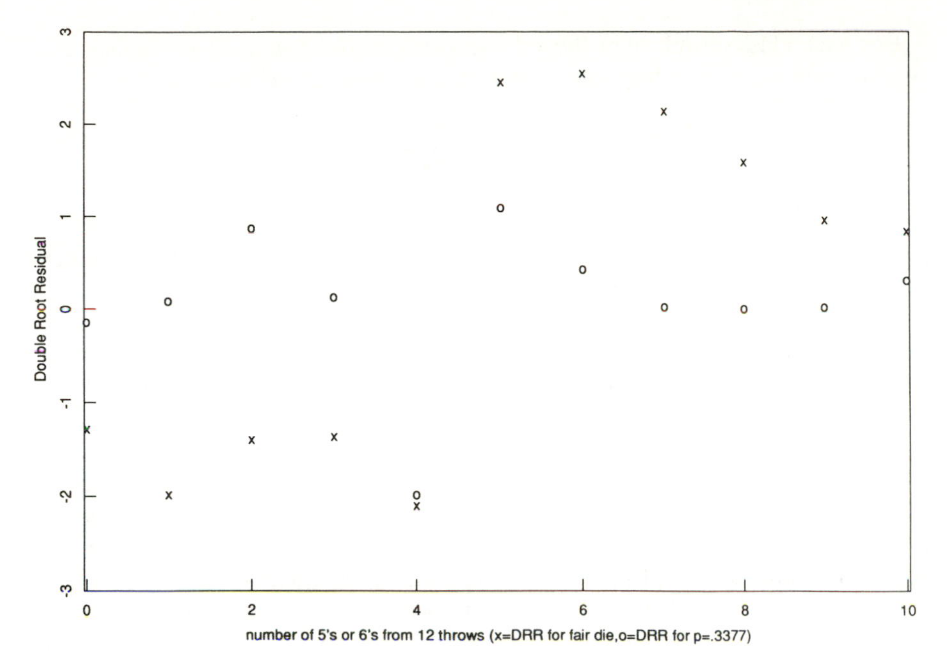

Figure 8.6 DRR plot of Weldon's dice data.

example. The DRRs for Weldon's dice data, both with the hypothesized and fitted p's are plotted in Figure 8.6. Clearly the DRRs discredit the binomial $N = 12$, $p = 1/3$ hypothesis, and confirm the fitted binomial hypothesis.

The analysis here has been most informative. The binomial model could fail for a variety of reasons—the trials might not have been independent, the probability p might not have been constant from trial to trial, and so on. Here, X_{PF}^2 confirms the binomial model, and the component V_1 of X_P^2 identifies that Np is not 4, thereby concluding that p is not 1/3.

8.4 Closing Remarks

Now that we have completed our investigation of the one sample goodness of fit problem, at least as far as we intend for the time being, it is appropriate to review that investigation.

Models have been divided into a double dichotomy of data categorized or uncategorized, and simple or composite null hypotheses. Distributions such as the binomial could be considered as either categorized or uncategorized, since there are finitely many values of the random variable or categories. Whether the null hypothesis is simple or composite depends on what is specified. Thus, there are ambiguities in our classification, but no inconsistencies result.

For each class of null hypotheses a smooth model has been proposed. For the uncategorized null hypotheses the distribution is specified by

$$g_k(x; \theta, \beta) = C(\theta, \beta) \exp\left\{\sum_{i=1}^{k} \theta_i h_i(x, \beta)\right\} f(x; \beta)$$

If there are no nuisance parameters, if $f(x; \beta)$ is the uniform $(0, 1)$ probability density function, and if the $h_i(x, \beta)$ are the normalized Legendre orthogonal polynomials, then the model is that given by Neyman more than 50 years ago. As former students of H. O. Lancaster, we have naturally looked at choosing $\{h_i(x, \beta)\}$ to be orthonormal on $f(x; \beta)$. We feel the simplicity and power of many of the tests that result have vindicated our preference.

This is not to say that those whose preference is to choose powers, $h_i(x, \beta) = x^i$, are "wrong." We are indebted to authors such as D.A. Pierce, D.R. Thomas, K.J. Kopecky, and J.A. Koziol, whose work largely precedes ours. Their tests are very similar to ours and, we believe, in the sense of Pitman's asymptotic relative efficiency, as efficient. Perhaps there is some choice of $h_i(x, \beta)$ that will prove in some sense to be superior to both schemes. That is not a problem. We are all trying to achieve the same thing: powerful, convenient, and informative model assessment.

For categorized data the smooth models are specified by the cell probabilities.

$$\pi_j(\beta) = C(\theta, \beta) \exp\left\{\sum_{i=1}^{k} \theta_i h_{ij}(\beta)\right\} p_j(\beta), \qquad j = 1, \ldots, m$$

When there are no nuisance parameters, Pearson's X^2 test follows from this model. When there are nuisance parameters, the Pearson–Fisher test based on X^2_{PF} and the Rao–Robson test based on X^2_{RR} follow. By choosing the $h_{ij}(\beta)$ appropriately, other tests may be obtained. In particular, we have investigated tests that overlap cells, and others that focus on particular (groupings and weightings of) cells. Here again, we must not be straight-jacketted into always choosing powers or orthonormal functions. Inventive choices may yield powerful new and useful tests.

The same comment applies to our specification of the $g_k(x; \theta, \beta)$ and $\pi_j(\beta)$. Perhaps some modification there will prove fruitful. Our choice has the advantage that the score tests can be fairly easily applied to the models and produce asymptotically optimal tests, and statistics with convenient χ^2 null and alternative distributions. Also, the tests based on the components are score tests with their "nice" properties. The components are very informative in applications, and their accessibility and convenience are especially welcome. In categorized situations, the components may prove to be less accessible than some class of contrasts; but the same properties and comments apply to these.

One aspect of the investigation that remains incomplete is the investigation of the small sample properties of smooth tests for particular uncategorized distributions. Of particular interest are the corrections necessary to the critical points to more nearly achieve the nominal significance level. Of the distributions yet to be addressed, the binomial is perhaps the greatest omission, followed by the gamma, negative binomial, and Weibull. We hope this monograph will stimulate these size and also power studies for distributions of interest to particular researchers. In particular cases the confirmatory studies may not be necessary. Example 8.3.1 was one such case.

In general, we claim that the use of components of the smooth tests overcomes one of the major criticisms of Pearson's X^2 test in particular, and goodness of fit tests in general—that when the null hypothesis is rejected there is no alternative model apparent. Component analysis permits a very close scrutiny of the model, and often leads to a much deeper understanding of the generating mechanism. The fact that our components are asymptotically independent (why use nonorthogonal contrasts?) is often an important part of this improved perception.

In particular cases, the smooth tests we have proposed are very competitive with existing tests of fit. In many cases, the first one or two components of our test statistics are already known and widely used goodness of fit tests. The first two components of the smooth test for the normal are standardized versions of the skewness and kurtosis statistics. For the exponential, our first component is a linear translation of the statistic given by Greenwood (1946). The binomial and Poisson first components are standardized versions of R. A. Fisher's indices of dispersion, while the first component of the smooth test for the geometric is clearly related to Vit's test.

There are several questions we have left unanswered. The confirmatory size and power studies already mentioned are of interest. Also of interest is the effect of using moment and other convenient estimators instead of the maximum likelihood estimators on the null distribution of the test statistic and the power of the test. How does the significance level vary with the nuisance parameter? How do our single sample results carry over to the general problem of k-sample goodness of fit? Smooth models, score tests, and overlapping are all relevant in contingency table analysis. The thoughts outlined in §8.2 are open to expansion and development. We hope to turn to some of these in the immediate future, and would be happy to hear from readers who are interested in joining these quests.

APPENDIX 1

Orthogonal Polynomials

In this Appendix we will give, either explicitly or in terms of their recurrence relations, some of the orthogonal polynomials used in this monograph. These relations provide a more numerically stable method for calculating orthogonal polynomials than, for example, explicit summation formulas.

First, we will give relations for the Meixner (1934) class. A *centered* random variable has zero mean; an *uncentered* random variable Y can be centered by defining $X = Y - E[Y]$. For the centered random variable X with orthogonal polynomials $\{P_n(x)\}$ the recurrence relation is

$$P_{n+1}(x) = (x + n\lambda)P_n(x) + n(-\sigma^2 + (n-1)k)P_{n-1}(x) \qquad \text{(A.2.1)}$$

for $n = 0, 1, 2, \ldots$, where we define $P_{-1}(x) = 0$, $P_1(x) = 1$ and $\sigma^2 = \text{var}(X)$. The constant λ can be determined from $P_2(0)$ and the constant k from $P_3(0)$; $P_2(0)$, and $P_3(0)$ are easily obtained from the summation formula or otherwise. Lancaster (1975) gave a proof of the Meixner relation (Equation A.2.1). To obtain orthonormal polynomials the orthogonal polynomials $P_n(x)$ must be normalized by dividing each by $\sqrt{E[\{P_n(x)\}^2]} = s_n$ say. This standardization factor can also be given recursively. Write $P_n^*(x) = P_n(x)/s_n$. The first five distributions in the Meixner family follow.

1. *Standard Normal.* The probability density function is $f(x) = \sqrt{(2\pi)} \exp(-x^2/2)$, $-\infty < x < \infty$, and Equation A.2.1 reduces to

   $$P_{n+1}(x) = xP_n(x) - nP_{n-1}(x)$$

   This yields

   $$P_n(x) = \sum_{t=0}^{[n/2]} \frac{(-1/2)^t n!}{t!\,(n-2t)!} x^{n-2t}$$

   The standardization factor *is* $s_n = \sqrt{(n!)}$.
2. *Poisson.* For $\lambda > 0$, the probability function is $f(x; \lambda) = e^{-\lambda} x^{\lambda}/x!$,

$x = 0, 1, 2, \ldots$. the standardizing factor is $s_n = \surd(\lambda n!)$, and the orthonormal polynomials are

$$P_n^*(x) = \surd\{\lambda^n/n!)\} \sum_{t=0}^{n} (-1)^{n-t}\, {}^xC_t t!\, \lambda^{-t}\, {}^nC_t$$

The following recurrence relation holds:

$$\surd(n\lambda)P_n^*(x) = \{(n + x - \lambda - 1)P_{n-1}^*(x-1) - \surd[(n-1)/\lambda](x-1)P_{n-2}^*(x-2)\},\ n \geq 2,\ x \geq 2$$

3. *Exponential.* The probability density function for the uncentered random variable Y is $f(y) = e^{-y}$, $y > 0$, zero otherwise. As $s_n = n!$,

$$P_n^*(y) = \sum_{t=0}^{n} {}^nC_t(-y)^t/t!$$

If Y is unscaled, so that the probability density function is $f(y) = \lambda e^{-\lambda y}$, $y > 0$, zero otherwise, then the recurrence relation is,

if $z = y/\bar{y}$, $P_1^*(z) = 1 - z$, $P_2^*(z) = 1 - 2z + z^2$, then for $n > 2$,

$$nP_n^*(z) = (2n - 1 - z)P_{n-1}^*(z) - (n-1)P_{n-2}^*(z)$$

4. *Binomial.* The probability function is ${}^NC_x p^x(1-p)^{N-x}$, $x = 0, 1, \ldots, N$. Write $q = 1 - p$. It can be shown that $s_n^2 = {}^NC_n(p/q)^n$. If we write $g = -\surd(p/q)$ and $h = \surd(q/p)$ then

$$\sqrt{\{{}^NC_n\}}P_n^*(x) = \sum_{t=0}^{n} {}^xC_t\, {}^{N-x}C_{n-t}\, g^{n-t}h^t$$

5. *Geometric.* The probability function is $f(x; q) = pq^x$, $x = 0, 1, \ldots,$ in which $q = 1 - p$. It can be shown that $s_n^2 = (n!)^2(q/p^2)^n$. Formulas for this distribution are given in §6.5.

   The discrete and continuous uniform distributions have also been referred to.

6. *Continuous uniform.* If the support is $(0, 1)$, the probability density function is $f(x) = 1$, $x \in (0, 1)$, zero otherwise. Now

if $z = 2x - 1$ and if $P_1^*(z) = z$ and $P_2^*(z) = (3z^2 - 1)/2$, then

$$nP_n^*(z) = (2n-1)zP_{n-1}^*(z) - (n-1)P_{n-2}^*(z),\ n \geq 2$$

7. *Discrete uniform.* Take the probability function to be $p_j = 1/m$, $j = 0, 1, \ldots, m - 1$. Now write $\alpha_n = n^2(m^2 - n^2)/\{4(4n^2 - 1)\}$. Then

if $P_0(j) = 1$ and $P_1(j) = j - (m-1)/2$,

$$P_{n+1}(j) = P_1(j)P_n(j) - \alpha_n P_{n-1}(j), \qquad n \geq 1$$

We also find $s_n^2 = \sum_{j=0}^{m-1} \{P_n(j)\}^2/m$.

APPENDIX 2

Computer Program for Implementing the Smooth Tests of Fit for the Uniform, Normal, and Exponential Distributions

Language

Fortran 77.

Description and Purpose

Given a sample of $n(\geq 10)$ independent observations $x_1, x_2, \ldots, x_n$ the following routine tests goodness of fit for any one of the standard uniform, normal, or exponential distributions. The test for uniformity is the smooth test based on the Legendre polynomials suggested by Neyman (1937) and discussed in Chapters 1 and 4. The normality and exponentiality tests are discussed in §§6.2 and 6.3. The appropriate orthogonal polynomials in these cases are the Hermite and the Laguerre. The test provided by the algorithm should be used in conjunction with a graphic technique such as a probability plot or histogram. For $n < 10$, where the approximate critical values may err, an approximate P value could be found, if necessary, by Monte Carlo methods. In the following we will write U_i for what is denoted by $\hat{V}_i$ in the rest of the book.

Numerical Method

Uniformity

The statistic calculated is $S = U_1^2 + \ldots + U_4^2$ where

$$U_r = \sqrt{[(2r+1)/n]} \sum_{j=1}^{n} \pi_r(Z_j)$$

and $\pi_1(z) = z$, $\pi_2(z) = (3z^2 - 1)/2$, $r\pi_r(z) = (2r-1)z\pi_{r-1}(z) - (r-1)\pi_{r-2}(z)$ for $r > 2$, with $z = (2x - 1)$. To a good approximation, when

$n \geq 10$, the upper 5% critical value of S is the χ_4^2 value given by $7.78(1 - 0.06/\sqrt{n})$. The U_i here and in the following are approximately standard $N(0, 1)$. Solomon and Stephens (1983) noted that U_1 and U_2 detect changes in mean and variance.

Normality

In this case $S = U_3^2 + \ldots + U_6^2$ where

$$U_r = \sum_{j=1}^{n} \pi_r(Z_j)/\sqrt{(nr!)}$$

and $\pi_1(z) = z$, $\pi_2(z) = z^2 - 1$, $\pi_r(z) = z\pi_{r-1}(z) - (r-1)\pi_{r-2}(z)$ for $r > 2$, with $z = (x - a)/b$, $a = \bar{x}$ and $b = \sum_{i=1}^{n} (x_i - \bar{x})^2/n$. For $n \geq 10$ adjusted χ_4^2 values for the 5% and 10% upper critical values were found to be $9.49(1 - 1.6/\sqrt{n})$ and $7.78(1 - 1.8/\sqrt{n})$. Table A.1.1 shows actual approximate tail areas based on 1,000 samples each of size n for the adjusted and unadjusted χ_4^2 critical values. As in §1.2, we note that U_3 and U_4 are standardized versions of the usual sample skewness and kurtosis coefficients.

Table A.1.1 Approximate tail areas for samples of size n based on adjusted and unadjusted (in parenthesis) χ_4^2 critical values

n	5%	10%
10	.05 (.00)	.12 (.01)
15	.05 (.02)	.09 (.02)
20	.05 (.01)	.09 (.02)
30	.05 (.03)	.08 (.04)
50	.05 (.03)	.10 (.06)
70	.06 (.04)	.10 (.06)
90	.06 (.05)	.09 (.06)
200	.05 (.05)	.09 (.06)

Exponentiality

The statistic is now defined by $S = U_2^2 + \ldots + U_5^2$ where

$$U_r = \sum_{j=1}^{n} \pi_r(Z_j)/\sqrt{n}$$

with, if we write $z = x/\bar{x}$, $\pi_1(z) = 1 - z$, $\pi_2(z) = 1 - 2z + z^2/2$, and $r\pi_r(z) = (2r - 1 - z)\pi_{r-1}(z) - (r-1)\pi_{r-2}(z)$ for $r > 2$. For $n \geq 10$, adjusted χ_4^2 values for the 5% and 10% upper critical values were found to be $9.49(1 - 1.5/\sqrt{n})$ and $7.78(1 - 1.8/\sqrt{n})$. The component U_2 is related to an exponentiality test discussed by Greenwood (1946).

FORTRAN code

```
subroutine gof(x,n,k,s,u,c,d,ifault)
dimension x(n), u(6), h(6), f(4), g(4)
data f/2.449490, 4.898979, 10.95445, 26.83282/
data g/1.732051, 2.236068, 2.645751, 3.0/
c
c goodness of fit for uniform, normal
 and exponential distributions
c
ifault=1
if(n.lt.10)return
ifault=0
do 1 j=1,6
u(j)=0
1 continue
e=sqrt(float(n))
s=0
c
c after initializing perform appropriate test
c
c
c first, uniformity test
c
if (k.eq.0) then
  do 4 i=1,n
    z=2*x(i)-1
  h(1)=z
  h(2)=(3*z*z-1)/2
    do 2 k=3,4
    h(k)=((2*k-1)*z*h(k-1)-(k-1)*h(k-2))/k
2 continue
  do 3 j=1, 4
  u(j)=u(j)+g(j)*h(j)/e
3 continue
4 continue
  c=9.49
  d=7.78*(1-0.06/e)
c
c normality test
c
else if (k.eq.1) then
  a=0
  do 5 i=1,n
  a=a+x(i)
5 continue
  a=a/n
  b=0
  do 6 i=1,n
  t=x(i)-a
```

```
6 b=b+t*t
  b=sqrt(b/n)
  do 9 i=1,n
  z=(x(i)-a)/b
  h(1)=z
  h(2)=z*z-1
    do 7 k=3,6
    h(k)=z*h(k-1)-(k-1)*h(k-2)
7 continue
    do 8 j=3,6
    u(j)=u(j)+h(j)/(e*f(j-2))
8 continue
9 continue
  c=9.49*(1-1.6/e)
  d=7.78*(1-1.8/e)
c
c exponentiality test
c
else if (k.eq.2) then
  a=0
  do 10 i=1,n
  a=a+x(i)
10 continue
  a=a/n
  do 13 i=1,n
  z=x(i)/a
  h(1)=1-z
  h(2)=1-2*z+z*z/2
    do 11 k=3,5
    h(k)=((2*k-1-z)*h(k-1)-(k-1)*h(k-2))/k
11 continue
  do 12 j=2,5
  u(j)=u(j)+h(j)/e
12 continue
13 continue
  c=9.49*(1-1.5/e)
  d=7.78*(1-1.8/e)
else
  ifault=2
  return
endif
c
c calculate s
c
do 14 j=1,6
s=s+u(j)*u(j)
14 continue
  return
end
```

Formal parameters

X	Real array	input:	the data array
N	Integer	input:	the number of data points
K	Integer	input:	a parameter to indicate what is tested for, equal to 0 for uniform 1 for normal 2 for exponential
S	Real	output:	the value of the test statistic
U	Real array	output:	array of components of S
C	Real	output:	upper 5% critical value of S
D	Real	output:	upper 10% critical value of S
IFAULT	Integer	output:	a fault indicator, equal to 1 if $N<10$ 2 if $K>2$ or $K<0$ 0 otherwise

Test Data

Suppose it is desired to test the adequacy of the fit of the log-normal distribution to the annual maximum 24-hourly rainfall for 47 years for Turramurra (a northern suburb of Sydney, Australia). The data (in millimeters) are:

1468, 909, 841, 475, 846, 452, 3830, 1397, 556, 978, 1715, 747, 909, 2002, 1331, 1227, 2543, 2649, 1781, 1717, 2718, 584, 1859, 1138, 2675, 1872, 1359, 1544, 1372, 1334, 955, 1849, 719, 1737, 1389, 681, 1565, 701, 994, 1188, 962, 1564, 1800, 580, 1106, 880, 850

After taking logarithms subroutine GOF may be used with $K=1$ to obtain $S=0.59$ ($U_3=0.01$, $U_4=-0.67$, $U_5=0.25$, $U_6=0.28$). The S value is well below the approximate 5% critical value of $C=7.3$ ($D=5.7$); therefore, the lognormal fit is tenable.

APPENDIX 3

Explicit Formulas for the Components V_1 and V_2 of X_P^2

The components V_1 and V_2 that arise in Chapter 5, the completely specified categorized case, are given by

$$V_r = \sum N_j g_r(j), \; r = 1,2$$

The orthogonal polynomials $g_1(j)$ and $g_2(j)$ can be defined by

$$g_1(j) = A\{(j-1) - S_1\} \quad \text{and} \quad g_2(j) = C\{(j-1)^2 - A^2Y(j-1) + Z\}$$

where

$$S_t = \sum_j (j-1)^t p_j,$$

$$A = (S_2 - S_1^2)^{-1/2},$$

$$Y = S_3 - S_1S_2,$$

$$Z = A^2YS_1 - S_2, \text{ and}$$

$$C = (S_4 + A^4Y^2S_2 + Z^2 - 2A^2S_3 + 2ZS_2 - 2A^2YZS_1)^{-1/2}.$$

References

Aggarwal, L. K. and Bajgier, S. M. (1987). Comparative performance of tests of normality in detecting mixtures of parallel regression lines. *Commun. Statist.-Theor. Meth.* 16:2541–2563.

Anderson, R. L. (1959). Use of contingency tables in the analysis of consumer preference studies. *Biometrics* 15:582–590.

Anderson, T. W. (1958). *An introduction to multivariate statistical analysis.* New York, Wiley.

Andrews, D. F. and Herzberg, A. M. (1985). *Data.* New York, Springer-Verlag.

Angus, J. E. (1982). Goodness-of-fit tests for exponentiality based on a loss-of-memory type functional equation. *J. Statist. Planning. Infer.* 6:241–251.

Atkinson, A. C. (1985). *Plots, transformations and regression,* New York: Oxford University Press.

Aroian, L. A. (1937). The type B Gram–Charlier series. *Ann. Math. Statist.* 8:183–192.

Bargal, A. I. (1986). Smooth tests of fit for censored gamma samples. *Commun. Statist.-Theor. Meth.* 15:537–549.

Bargal, A. I. and Thomas, D. R. (1983). Smooth goodness of fit tests for the Weibull distribution with singly censored data. *Commun. Statist.-Theor. Meth.* 12:1431–1447.

Barnett, V. and Lewis, T. (1984). *Outliers in statistical data* (2nd ed.). New York, Wiley.

Barton, D. E. (1953). On Neyman's smooth test of goodness of fit and its power with respect to a particular system of alternatives. *Skand Aktuarietidskr* 36:24–63.

Barton, D. E. (1955). A form of Neyman's Ψ^2 test of goodness of fit applicable to grouped and discrete data. *Skand. Aktuarietidskr.* 38:1–16.

Barton, D. E. (1956). Neyman's Ψ^2 test of goodness of fit when the null hypothesis is composite. *Skand. Aktuarietidskr* 39:216–245.

Bennett, B. M. and Birch, J. B. (1974). On the small sample distribution and power of the log likelihood ratio and variance tests for the Poisson. *J. Statist. Comp. Simul.* 3:33–40.

Bera, A. and John, S. (1983). Tests for multivariate normality with Pearson alternatives. *Commun. Statist.-Theor. Meth.* 12:103–117.

Bera, A. K. and McKenzie, C. R. (1986). Tests for normality with stable alternatives. *J. Statist. Comp. Simul.* 25:37–52.

Best, D. J. (1979). Some easily programmed pseudo-random normal generators. *Austral. Comp. J.* 11:60–62.

Best, D. J. and Rayner, J. C. W. (1981). Are two classes enough for the X^2 goodness of fit test? *Statist. Neerl.* 35:157–163.

Best, D. J. and Rayner, J. C. W. (1982). Partitioning the equiprobable X^2 statistic for testing uniformity. *N. Z. Statistician* 17:29–32.

Best, D. J. and Rayner, J. C. W. (1985a). Uniformity testing when alternatives have low order. *Sankhya* A, 47:25–35.

Best, D. J. and Rayner, J. C. W. (1985b). Lancaster's test of normality. *J. Statist. Planning Infer.* 12:395–400.

Best, D. J. and Rayner, J. C. W. (1987a). Goodness-of-fit for grouped data using components of Pearson's X^2. *Comp. Statist. & Data Anal.* 5:53–57.

Best, D. J. and Rayner, J. C. W. (1987b). Welch's approximate solution for the Behrens–Fisher problem. *Technometrics* 29:205–210.

Best, D. J. and Rayner, J. C. W. (1988). A test for bivariate normality. *Statist. Prob. Letters,* 6:407–412.

Best, D. J., Rayner, J. C. W., and Turnbull, A. P. (1983). Power approximations for Pearson's chi-squared test. *Statcomp '83' papers.* Statist. Soc. Austral., 41–49.

Bhat, B. R. and Nagnur, B. N. (1965). Locally asymptotically most stringent tests and Lagrangian multiplier tests of linear hypotheses. *Biometrika* 52:459–468.

Bozdogan, H. and Ramirez, D. E. (1986). An adjusted likelihood-ratio approach to the Behrens–Fisher problem. *Commun. Statist.-Theor. Meth.* 15:2405–2433.

Buckley, M. (1987). Testing normality of residuals. Consulting Report, NSW87/35/MJB13, Division of Mathematics and Statistics, CSIRO, Sydney, Australia.

Buse, A. (1982). The likelihood-ratio, Wald, and Lagrange multiplier tests: an expository note. *Amer. Statistician* 36:153–157.

Campbell, D. B. and Oprian, C. A. (1979). On the Kolmogorov–Smirnov test for the Poisson distribution with unknown mean. *Biom. J.* 21:17–24.

Chacko, V. J. (1966). Modified chi-square test for ordered alternatives. *Sankhya* B, 28:185–190.

Chambers, J. M., Cleveland, W. S., Kleiner, B., and Tukey, P. A. (1983). *Graphical methods for data analysis,* Boston: Duxbury Press.

Chandra, T. K. and Joshi, S. N. (1983). Comparison of the likelihood ratio, Rao's and Wald's test and a conjecture of C. R. Rao. *Sankhya* A, 45:226–246.

Chernoff, H. and Lehmann, E. L. (1954). The use of maximum likelihood estimates in χ^2 tests for goodness of fit. *Ann. Math. Statist.* 25:579–586.

Cochran, W. G. (1952). The χ^2 test of goodness of fit. *Ann. Math. Statist.* 23:315–345.

Cochran, W. G. (1954). Some methods of strengthening the common χ^2 tests. *Biometrics* 10:417–451.

Cohen, A. and Sackrowitz,. H. B. (1975). Unbiasedness of the chi-squared, likelihood ratio, and other goodness of fit tests for the equal cell case. *Ann. Statist.* 3:959–964.

Conover, W. J. (1980). *Practical nonparametric statistics.* New York: Wiley.

Cox, D. R. (1977). Discussion of "Do robust estimators work with real data?". *Ann. Statist.* 5:1083.

Cox, D. R. and Hinkley, D. V. (1974). *Theoretical statistics*. London: Chapman and Hall.

Cox, D. R. and Small, N. J. H. (1978). Testing multivariate normality. *Biometrika* 65:263–272.

Cramer, H. (1946). *Mathematical methods of statistics*. Princeton: Princeton University Press.

Cressie, N. and Read, T. R. C. (1984). Multinomial goodness-of-fit tests. *J. R. Statist. Soc.* B, 46:440–464.

Csorgo, S. (1986). Testing for normality in arbitrary dimension. *Ann. Statist.* 14:708–723.

D'Agostino, R. B. and Stephens, M. A. (1986). *Goodness-of-fit techniques*. New York: Marcel Dekker.

Dahiya, R. C. and Gurland, J. (1972). Pearson chi-square test of fit with random intervals. *Biometrika* 59:147–153.

Dahiya, R. C. and Gurland, J. (1973). How many classes in the Pearson chi-square test? *J. Amer. Statist. Ass.* 68:707–712.

David, H. J. (1966). Goodness of fit. In *The Encyclopedia of Statistics*. Vol. 1 (editors Kruskal, W. H. and Tanur, J. M.), pp. 399–409. New York: The Free Press.

Devroye, L. (1986). *Non-uniform random variate generation*. New York: Springer-Verlag.

Diggle, P. J. and Hall, P. (1986). The selection of terms in an orthogonal series density estimator. *J. Amer. Statist. Ass.* 81:230–233.

Durbin, J. and Knott, M. (1972). Components of Cramer-von Mises statistics I, *J. R. Statist. Soc.* B, 34:290–307.

Edwards, J. H. (1961). The recognition and estimation of cyclic trends. *Ann. Hum. Gen.* 25:83–86.

Elderton, W. A. and Johnson, N. L. (1969). *Systems of frequency curves*. Cambridge: Cambridge University Press.

Eubank, R. L., LaRiccia, V. N., and Rosenstein, R. B. (1987). Test statistics derived as components of Pearson's Phi-squared distance measure. *J. Amer. Statist. Ass.* 82:816–825.

Filliben, J. F. (1975). The probability plot correlation coefficient test for normality. *Technometrics* 17:111–117.

Fisher, R. A. (1925). *Statistical methods for research workers* (14th ed. 1970). Edinburgh: Oliver and Boyd.

Fisher, R. A. and Yates, F. (1963). *Statistical tables for biological, agricultural and medical research* (6th ed.). Edinburgh: Oliver and Boyd.

Freedman, L. S. (1981). Watson's U_N^2 statistic for a discrete distribution. *Biometrika* 68:708–711.

Gail, M. H. and Gastwirth, J. L. (1978a). A scale-free goodness-of-fit test for the exponential distribution based on the Gini statistic. *J. R. Statist. Soc.* B, 40:350–357.

Gail, M. H. and Gastwirth, J. L. (1978b). A scale-free goodness-of-fit test for the exponential distribution based on the Lorenz curve. *J. Amer. Statist. Ass.* 73:787–793.

Galton, F. (1888). Co-relations and their measurements, chiefly from Anthropometric data. *Proc. Royal Soc* 45:135–140.

Gart, J. J. (1975). The Poisson distribution: the theory and application of some conditional tests. *A Modern Course on Statistical Distributions in Scientific Work* 2:125–140.

Gart, J. J. and Tarone, R. E. (1983). The relation between score tests and approximate UMPU tests in exponential models common in biometry. *Biometrics* 39:781–786.

Gbur, E. E. (1981). On the Poisson index of dispersion. *Commun. Statist.-Simul. Comp.* 10:5315–535.

Geary, R. C. (1947). Testing for normality. *Biometrika* 34:209–242.

Gleser, L. J. and Moore, D. S. (1985). The effect of positive dependence on chi-squared tests for categorical data. *J. R. Statist. Soc.* B, 47:459–465.

Gnanadesikan, R. (1977). *Methods for statistical analysis of multivariate observations.* New York: Wiley.

Greenwood, M. (1946). The statistical study of infectious diseases. *J. R. Statist. Soc.* 109:85–110.

Gumbel, E. J. (1943). On the reliability of the classical χ^2 test. *Ann. Math. Statist.* 14:253–263.

Haight, F. A. (1967). *Handbook of the Poisson distribution.* New York: Wiley.

Hall, P. (1985). Tailor-made tests of goodness of fit. *J. R. Statist. Soc.* B, 47:125–131.

Hamdan, M. A. (1962). The powers of certain smooth tests of goodness of fit. *Aust. J. Statist.* 4:25–40.

Hamdan, M. A. (1963). The number and width of classes in the chi-square test. *J. Amer. Statist. Ass.* 58:678–689.

Hamdan, M. A. (1964). A smooth test of goodness of fit based on the Walsh functions. *Austral. J. Statist.* 6:130–136.

Harrison, R. H. (1985). Choosing the optimum number of classes in the chi-square test for arbitrary power levels. *Sankhya* B, 47:319–324.

Hauck, W. W. and Donner, A. (1977). Wald's test as applied to hypotheses in logit analysis. *J. Amer. Statist. Ass.* 72:851–853. Corrigendum (1980), 75, 482.

Hirotsu, C. (1986). Cumulative chi-squared statistic as a tool for testing goodness of fit. *Biometrika* 73:165–173.

Hoaglin, D. C. (1980). A Poissonness plot. *Amer. Statistician* 34:146–149.

Hoaglin, D. C., Mostellar, F., and Tukey, J. W. (1985). *Exploring data tables, trends and shapes.* New York: Wiley.

Hogg, R. V. (1978). *Studies in statistics.* Washington: Math. Assoc. of Amer.

Holtzman, G. I. and Good, I. J. (1986). The Poisson and chi-squared approximations as compared with the true upper-tail probability of Pearson's X^2 for equiprobable multinomials. *J. Statist. Planning Infer.* 13:283–295.

Horn, S. D. (1977). Goodness-of-fit tests for discrete data: a review and an application to a health impairment scale. *Biometrics* 33:237–248.

Huang, C. J. and Bolch, B. W. (1974). On the testing of regression disturbances for normality. *J. Amer. Statist. Ass.* 69:330–335.

Hutchinson, T. P. (1979). The validity of the chi-square test when expected frequencies are small: a list of recent research references. *Commun. Statist.-Theor. Meth.* 8:327–335.

IMSL Library (1982). *Reference Manual* (9th ed.). Houston: IMSL.

Jacque, C. M. and Bera, A. K. (1987). A test for normality of observations and regression residuals. *Int. Statist. Rev.* 55:163–177.

Javitz, H. S. (1975). Generalized smooth tests of goodness of fit, independence, and equality of distributions. Unpublished thesis, Univ. of Calif., Berkeley.

Johnson, N. L. and Kotz, S. (1969). *Distributions in statistics: discrete distributions.* Boston: Houghton Mifflin.

Kallenberg, W. C. M. (1985). On moderate and large deviations in multinomial distributions. *Ann. Math. Statist.* 13:1554–1580.

Kallenberg, W. C. M., Oosterhoff, J., and Schriever, B. F. (1985). The number of classes in chi-squared goodness-of-fit tests. *J. Amer. Statist. Ass.* 80:959–968.

Kang, Soo-Il (1979). Performance of generalized Neyman smooth goodness of fit tests. Unpublished Ph.D. thesis, Department of Statistics, Oregon State University.

Katti, S. K. (1973). Exact distribution for the chi-squared test in the one way table. *Commun. Statist.-Theor. Meth.* 2:435–447.

Kempthorne, O. (1966). The classical problem of inference-goodness of fit. In *Proceedings of the Fifth Berkeley Symposium on Mathematical Statistics and Probability,* Vol. I (Neyman, J. ed.), pp. 235–249. Berkeley: Univ. of Calif. Press.

Kendall, M. G. and Stuart, A. S. (1973). *The advanced theory of statistics.* Vol. 2 (3rd ed.). London: Griffin.

Kendall, M. G. and Stuart, A. S. (1977). *The advanced theory of statistics.* Vol. 1 (4th ed.). London: Griffin.

Kinderman, A. J. and Monahan, J. F. (1977). Computer generation of random variables using the ratio of uniform deviates. *ACM Trans. Maths. Software* 3:257–260.

Kitagawa, G. (1979). On the use of AIC for the detection of outliers. *Technometrics* 21:193–199.

Koehler, K. J. (1979). A general formula for moments of the Pearson goodness-of-fit statistic for alternatives. *Biometrika* 66:397–399.

Koehler, K. J. and Larntz, K. (1980). An empirical investigation of goodness-of-fit statistics for sparse multinomials. *J. Amer. Statist. Ass.* 75:336–344.

Kopecky, K. J. and Pierce, D. A. (1979). Efficiency of smooth goodness-of-fit tests. *J. Amer. Statist. Ass.* 74:393–397.

Koziol, J. A. (1986). Assessing multivariate normality: a compendium. *Commun. Statist.-Theor. Meth.* 15:2763–2783.

Koziol, J. A. (1987). An alternative formulation of Neyman's smooth goodness of fit tests under composite alternatives. *Metrika* 34:17–24.

Lancaster, H. O. (1958). The structure of bivariate distributions. *Ann. Math. Statist.* 29:719–736.

Lancaster, H. O. (1965). The Helmert matrices. *American Mathematical Monthly* 72:4–12.

Lancaster, H. O. (1969). *The chi-squared distribution.* New York: Wiley.

Lancaster, H. O. (1975). Joint probability distributions in the Meixner classes. *J. R. Statist. Soc.* B, 37:434–443.

Lancaster, H. O. (1980). Orthogonal models in contingency tables. In *Developments in Statistics,* Vol. 3 (Krishnaiah, P. R. ed.), pp. 99–157. New York: Academic Press.

Larntz, K. (1978). Small-sample comparisons of exact levels for chi-squared goodness-of-fit statistics. *J. Amer. Statist. Ass.* 73:253–263.

Larsen, R. J. and Marx, M. L. (1981). *An introduction to mathematical statistics and its applications.* Englewood Cliffs: Prentice-Hall.

Lawal, H. B. (1980). Tables of percentage points of Pearson's goodness-of-fit statistic for use with small expectations. *Appl. Statist.* 29:292–298.

Lawless, J. F. (1982). *Statistical models and methods for lifetime data.* New York: Wiley.

Lee, C. C. (1987). Chi-squared tests for and against an order restriction on multinomial parameters. *J. Amer. Statist. Ass.* 82:611–618.

Lehmann, E. L. (1986). *Testing statistical hypotheses* (2nd ed.). New York: Wiley.

Lin, C. C. and Mudholkar, G. S. (1980). A test of exponentiality based on the bivariate F distribution. *Technometrics* 22:79–82.

Machado, S. G. (1983). Two statistics for testing multivariate normality. *Biometrika* 70:713–718.

Mahalanobis, P. C. (1934). A revision of Risley's anthropometric data relating to the Chittagong hill tribes. *Sankhya* B, 1:267–276.

Malkovich, J. F. and Afifi, A. A. (1973). On tests for multivariate normality. *J. Amer. Statist. Ass.* 68:176–179.

Mann, H. B. and Wald, A. (1942). On the choice of the number of intervals in the application of the chi-squared test. *Ann. Math. Statist.* 13:306–317.

Mantel, N. (1987). Understanding Wald's test for exponential families. *Amer. Statistician* 41:147–148.

Mardia, K. V. (1970). Measures of multivariate skewness and kurtosis with applications. *Biometrika* 57:519–530.

Mardia, K. V. (1974). Applications of some measures of multivariate skewness and kurtosis in testing normality and robustness studies. *Sankhya* A, 36:115–128.

Mardia, K. V. (1980). Tests of univariate and multivariate normality. In *Handbook of Statistics.* Vol. 1 (Krishnaiah, P. R. ed.), pp. 279–320. New York: North-Holland.

Mardia, K. V. (1986). Mardia's test of multinormality. In *Encyclopedia of Statistics,* Vol. 5 (Johnson, N. L., Kotz, S., and Read, C. eds.), pp. 217–221. New York: Wiley.

Mardia, K. V. and Foster, K. (1983). Omnibus tests of multinormality based on skewness and kurtosis. *Commun. Statist.-Theor. Meth.* 12:207–221.

Meixner, J. (1934). Orthogonale polynomsysteme mit einer besondenen Gestalt der erzeugenden Funktion. *J. Lond. Math. Soc.* 9:6–13.

Michael, J. R. (1983). The stabilized probability plot. *Biometrika* 70:11–17.

Miller, F. L. and Quesenberry, C. P. (1979). Power studies of some tests for uniformity II. *Commun. Statist.-Simul. Comp.* 8:271–290.

Mood, A. M., Graybill, F. A., and Boes, D. C. (1974). *Introduction to the theory of statistics* (3rd ed.). Tokyo: McGraw-Hill Kogakusha,

Moore, D. S. (1977). Generalized inverses, Wald's method, and the construction of the chi-squared tests of fit. *J. Amer. Statist. Ass.* 72:131–137.

Moore, D. S. (1986). Tests of chi-squared type. In *Goodness-of-fit Techniques* (D'Agostino, R. B. and Stephens, M. A., ed.), pp. 63–95, New York: Marcel Dekker.

Moore, D. S. and Spruill, M. C. (1975). Unified large-sample theory of general chi-squared statistics for tests of fit. *Ann. Statist.* 3:599–616.

Moran, P. A. P. (1973). Asymptotic properties of homogeneity tests. *Biometrika* 60:79–85.

Neyman, J. (1937). "Smooth" test for goodness of fit. *Skand. Aktuarietidskr.* 20:150–199.

Neyman, J. and Pearson, E. S. (1928). On the use and interpretation of certain test criteria for purposes of statistical inference. *Biometrika* 20:175–240 and 263–294.

Neyman, J. and Pearson, E. S. (1931). Further notes on the χ^2 distribution. *Biometrika* 22:298–305.

Oosterhoff, J. (1985). The choice of cells in chi-square tests. *Statist. Neerl.* 39:115–128.

Pearson, E. S. (1938). The probability integral transformation for testing goodness-of-fit and combining independent tests of significance. *Biometrika* 30:134–148.

Pearson, E. S. (1956). Some aspects of the geometry of statistics. *J. R. Statist. Soc.* A, 119:125–146.

Pearson, E. S., D'Agostino, R. B., and Bowman, K. C. (1977). Tests for departure from normality: comparison of powers. *Biometrika* 64:231–246.

Pearson, E. S. and Hartley, H. O. (1970). *Biometrika tables for statisticians*. Vol. 1 (3rd ed.). New York: Cambridge University Press.

Pearson, E. S. and Hartley, H. O. (1972). *Biometrika tables for statisticians*. Vol. 2. New York: Cambridge University Press.

Pearson, K. (1897). Cloudiness: note on a novel case of frequency. *Proc. Royal Soc.* 62:287–290.

Pearson, K. (1900). On the criterion that a given system of deviations from the probable in the case of a correlated system of variables is such that it can reasonably be supposed to have arisen from random sampling. *Philos. Mag.*, 5th ser., 50:157–175.

Pearson, K. (1901). *Philos. Mag.* 6th ser., 1:670–671. (An untitled letter to the editor.)

Pearson, K. and Lee, A. (1903). On the laws of inheritance in man. I. Inheritance of physical characters. *Biometrika* 2:357–462.

Pettitt, A. N. (1977). Testing the normality of several independent samples using the Anderson–Darling statistic. *Appl. Statist.* 26:156–161.

Pettitt, A. N. (1979). Testing for bivariate normality using the empirical distribution function. *Commun. Statist.-Theor. Meth.* 8:699–712.

Pettitt, A. N. and Stephens, M. A. (1977). The Kolmogorov–Smirnov goodness-of-fit statistic with discrete and grouped data. *Technometrics* 19:205–210.

Pierce, D. A. and Gray, R. G. (1982). Testing normality of errors in regression models. *Biometrika* 64:233–236.

Plackett, R. L. (1983). Karl Pearson and the chi-squared test. *Int. Statist. Rev.* 51:59–72.

Prescott, P. (1976). Comparison of tests for normality using stylized surfaces. *Biometrika* 63:285–289.

Quesenberry, C. P., Giesbrecht, F. G., and Burns, J. C. (1983). Some methods for studying the validity of normal model assumptions for multiple samples. *Biometrics* 39:735–739.

Quesenberry, C. P. and Hales, C. (1980). Concentration bands for uniformity plots. *J. Statist. Comp. Simul.* 11:41–53.

Quesenberry, C. P. and Miller, F. L. (1977). Power studies of some tests for uniformity. *J. Statist. Comp. Simul.* 5:169–191.

Radlow, R. and Alf, E. F. (1975). An alternative multinomial assessment of the accuracy of the χ^2 test of goodness of fit. *J. Amer. Statist. Ass.* 70:811–813.

Rao, C. R. (1948). Tests of significance in multivariate analysis. *Biometrika* 35:58–79.

Rao, K. C. and Robson, D. S. (1974). A chi-square statistic for goodness-of-fit tests within the exponential family. *Comm. Statist.* 3:1139–1153.

Rayner, J. C. W. and Best, D. J. (1982). The choice of class probabilities and number of classes for the simple χ^2 goodness of fit test. *Sankhya* B, 44:28–38.

Rayner, J. C. W. and Best, D. J. (1986). Neyman-type smooth tests for location-scale families. *Biometrika* 73:437–446.

Rayner, J. C. W., Best, D. J., and Dodds, K. G. (1985). The construction of the simple X^2 and Neyman smooth goodness of fit tests. *Statist. Neerl.* 39:35–50.

Rayner, J. C. W. and McIntyre, R. I. (1985). Use of the score statistic for testing goodness of fit of some generalised distributions. *Biom. J.* 27:159–166.

Read, T. R. C. (1984a). Small sample comparisons for the power divergence goodness-of-fit statistics. *J. Amer. Statist. Ass.* 79:929–935.

Read, T. R. C. (1984b). Closer asymptotic approximations for the distributions of the power of divergence goodness-of-fit statistics. *Ann. Inst. Statist. Math.* 36:59–69.

Risebrough, R. W. (1972). Effects of environmental pollutants upon animals other than man. *Proceedings of the 6th Berkeley Symposium on Mathematics and Statistics VI,* Berkeley: Univ. of Calif. Press, pp. 443–463.

Ronchetti, E. (1987). Robust $C(\alpha)$-type tests for linear models. *Sankhya* A, 49:1–16.

Roscoe, J. T. and Byars, J. A. (1971). An investigation of the restraints with respect to sample size commonly imposed on the use of the chi-square statistic. *J. Amer. Statist. Ass.* 66:755–759.

Roy, A. R. (1956). On χ^2 statistics with variable intervals. Technical report, Stanford University Statistics Department.

Scariano, S. M. and Davenport, J. M. (1986). A four-moment approach and other practical solutions to the Behrens–Fisher problem. *Commun. Statist.-Theor. Meth.* 15:1467–1505.

Schemper, M. (1987). Nonparametric estimation of variance, skewness and kurtosis of the distribution of a statistic by jackknife and bootstrap techniques. *Statist. Neerl.* 41:49–64.

Schorr, B. (1974). On the choice of the class intervals in the application of the chi-squared test. *Math. Operationsforsch. Statist.* 5:357–377.

Selby, B. (1965). The index of dispersion as a test statistic. *Biometrika* 52, 627–629.

Seyb, A. (1984). Comparison of some smooth goodness of fit tests. Unpublished Postgraduate Diploma project, Department of Mathematics and Statistics, University of Otago.

Shapiro, S. S. and Gross, A. J. (1981). *Statistical modelling techniques.* New York: Marcel Dekker.

Shirahata, S. (1987). A goodness of fit test based on some graphical representation when parameters are estimated. *Comp. Statist & Data Anal.* 5:127–136.

Silverman, B. W. (1986). *Density estimation for statistics and data analysis.* London: Chapman Hall.

Singh, A. C. (1987). On the optimality and a generalization of Rao–Robson's statistic. *Commun. Statist.-Theor. Meth* 16:3255–3273.

Smith, P. J., Rae, D. S., Manderscheid, R. W. and Silbergold, S. (1979). Exact and approximate distributions of the chi-square statistic for equiprobability. *Commun. Statist.-Simul. Comp.* 8:131-149.

Snedecor, G. W. and Cochran, W. G. (1980). *Statistical methods.* Ames, Iowa: Iowa State University Press.

Solomon, H. and Stephens, M. A. (1983). On Neyman's statistics for testing uniformity. *Commun. Statist.-Simul. Comp* 12:127–134.

Spruill, M. C. (1976). Cell selection in the Chernoff-Lehmann chi-square statistic. *Ann. Statist.* 4:375–383.

Srivastava, M. S. and Hui, T. K. (1987). On assessing multivariate normality based on Shapiro-Wilk W statistic. *Statist. Prob. Letters* 5:15–18.

Stephens, M. A. (1966). Statistics connected with the uniform distribution: percentage points and application to tests for randomness of directions. *Biometrika* 53:235–240.

Stephens, M. A. (1974a). EDF statistics for goodness of fit and some comparisons. *J. Amer. Statist. Ass.* 69:730–737

Stephens, M. A. (1974b). Components of goodness-of-fit statistics. *Ann. Inst. Henri Poincare,* 10:37–54..

Svensson, A. (1985). On χ^2 test of goodness-of-fit for a class of discrete multivariate models. *Statist. Prob. Letters* 3:331–336.

Szego, G. (1959). *Orthogonal Polynomials.* Colloquium Publications No. 23, New York: Amer. Math. Soc.

Tate, M. W. and Hyer, L. A. (1973). Inaccuracy of the χ^2 test of goodness of fit when expected frequencies are small. *J. Amer. Statist. Ass.* 68:836–841.

Thomas, D. R. and Pierce, D. A. (1979). Neyman's smooth goodness-of-fit test when the hypothesis is composite. *J. Amer. Statist. Ass.* 74:441–445.

Tiku, M. L., Tan, W. Y., and Balakrishnan, N. (1986). *Robust inference.* New York: Marcel Dekker.

Vaeth, M. (1985). On the use of Wald's test in exponential families. *Int. Statist. Rev.* 53:199–214.

Vit, P. (1974). Testing for homogeneity; the geometric distribution. *Biometrika* 61:565–568.

Wald, A. (1943). Tests of statistical hypotheses concerning several parameters when the number of observations is large. *Trans. Amer. Math. Soc.* 54:426–482.

Watson, G. S. (1959). Some recent results in chi-square goodness-of-fit tests. *Biometrics* 15:440–468.

Weisberg, S. (1980). *Applied linear regression.* New York: Wiley.

Welch, B. L. (1937). The significance of the difference between two means when the population variances are unequal. *Biometrika* 29:350–362.

West, E. N. and Kempthorne, O. (1971). A comparison of the chi^2 and likelihood ratio tests for composite alternatives. *J. Statist. Comp. Simul.* 1:1–33.

Wetherill, G. B. (1967). *Elementary statistical methods.* London: Chapman and Hall.

Wetherill, G. B. (1986). *Regression Analysis with Applications.* London: Chapman and Hall.

Wilk, M. B. and Gnanadesikan, R. (1968). Probability plotting methods for the analysis of data. *Biometrika* 55:1–17.

Williams, C. A. (1950). On the choice of the number and width of classes for the chi-square test of goodness of fit. *J. Amer. Statist. Ass.* 45:77–86.

Subject Index

Papers Index